常见乳畜的
乳特征性成分研究

马露 著

中国农业科学技术出版社

图书在版编目（CIP）数据

常见乳畜的乳特征性成分研究 / 马露著 . —北京：中国农业科学技术出版社，2018. 6

ISBN 978-7-5116-3693-5

Ⅰ. ①常… Ⅱ. ①马… Ⅲ. ①乳制品-研究 Ⅳ. ①TS252. 5

中国版本图书馆 CIP 数据核字（2018）第 092519 号

责任编辑 崔改泵 金 迪
责任校对 贾海霞

出 版 者 中国农业科学技术出版社
北京市中关村南大街 12 号 邮编：100081
电 话 (010)82109194(编辑室) (010)82109702(发行部)
(010)82109709(读者服务部)
传 真 (010)82106650
网 址 http://www.castp.cn
经 销 者 各地新华书店
印 刷 者 北京建宏印刷有限公司
开 本 710mm×1 000mm 1/16
印 张 8. 5
字 数 136 千字
版 次 2018 年 6 月第 1 版 2018 年 6 月第 1 次印刷
定 价 60. 00 元

前　言

乳营养丰富，是人类膳食结构中营养结构比较完善的一种食物，富含人类及新生儿所需的各种重要营养物质，包括氨基酸、维生素、矿物质等，参与人类大脑、免疫、神经和骨骼等系统的发育过程。乳蛋白富含必需氨基酸，具有较高的生物学价值，其中的乳清蛋白在一定程度上能够促进胰岛素的发挥作用，并可通过中短链脂肪酸发挥作用进而促进脂肪代谢。乳脂肪除可为机体提供能量外，还具有一些生理活性，乳脂中的中短链脂肪酸及不饱和脂肪酸均有益于人类健康，共轭亚油酸具有抗癌、抗炎、减少动脉粥样硬化和降低血脂的功效，除乳蛋白和乳脂肪外，乳中还富含有大量的维生素和矿物质。然而，不同畜种乳的潜在的营养价值还没有被充分地发掘出来。

本书系统地介绍了常规乳畜的乳特征性成分，以奶牛乳、水牛乳、牦牛乳、娟姗牛乳、山羊乳、骆驼乳和马乳为研究对象，利用质谱技术结合蛋白质组学技术，构建了奶牛乳、水牛乳、牦牛乳、娟姗牛乳、山羊乳、骆驼乳和马乳中乳蛋白及 OBCFA 的特征性图谱，确定了不同物种之间乳蛋白及 OBCFA 的特征性成分及其含量差异，并首次通过转录组学方法分析获得了新鲜奶牛与新鲜山羊乳清中特征性 microRNA 及反复冻融乳清与新鲜乳清样 microRNA 的表达差异性，发现了种属特异性以及对奶样冻融敏感的 microRNAs。通过对不同畜种乳特征乳成分的分析，揭示不同畜种乳特征差异性，为进一步阐释不同畜种乳营养成分及潜在的生物学功能的差异性，及评定乳品质提供一定的理论依据和基础信息，并为进一步区分不同畜种乳提供理论依据。

全书共分为 8 章，内容主要包括不同动物乳成分及其特征的研究进展、基于 RP-HPLC 技术分析不同乳畜乳中主要乳蛋白组分、基于 GC-MS 技术分析不同乳畜乳中 OBCFA 及其组分含量、基于 ICP-MS 方法分析不同畜种乳中常量及微量元素含量、基于 Label-Free 定量蛋白质组学方法分析不同畜种乳 MFGM 蛋白

以及基于转录组学方法分析新鲜奶牛与新鲜山羊乳清中特征性 microRNA 及反复冻融奶牛乳清与新鲜奶牛乳清样 microRNA 的差异性。

本书涉及的相关课题研究是在中国农业科学院北京畜牧兽医研究所反刍动物营养研究室完成的，也受到了内蒙古农业大学动物科学学院的大力支持。本书由国家奶业“973 计划”项目课题（编号：2011CB100805）、“十二五”国家科技支撑计划课题（编号：2010BAD12B02）以及中国农业科学院科技创新工程项目资助（编号：ASTIP-IAS07）。

由于作者水平有限，书中疏漏或不妥之处在所难免，恳请同行和读者批评指正。

作者

2018 年 3 月

目　录

1 不同乳畜乳成分及其特征的研究进展

乳营养丰富，是人类膳食结构中营养结构比较完善的一种食物，富含人类及新生儿所需的各种重要营养物质，包括氨基酸、维生素、矿物质等，参与人类大脑、免疫、神经和骨骼等系统的发育过程[1]。而草食性动物可以将自然界较为丰富但不易消化的纤维类物质消化发酵进而形成挥发酸、多肽、氨基酸和微生物等，然后经由乳腺合成分泌形成乳。乳蛋白富含必需氨基酸，具有较高的生物学价值。其中的乳清蛋白在一定程度上能够促进胰岛素发挥作用[2]，并可通过中短链脂肪酸发挥作用进而促进脂肪代谢[3]。乳脂肪除可为机体提供能量外，还具有一些生理活性，研究指出，乳脂中的中短链脂肪酸及不饱和脂肪酸均有益于人类健康[4]；此外，也有研究报道，共轭亚油酸具有抗癌、抗炎、减少动脉粥样硬化和降低血脂的功效[1,5]。除乳蛋白和乳脂肪外，乳中还富含有大量的维生素和矿物质。

1.1 国内外的研究进展

1.1.1 动物的乳产量及分布

目前全球食谱复杂多样，食品安全、营养的平衡以及疾病的流行等不容乐观。在发展中国家，乳畜是小规模牲畜养殖户保证家庭食物安全的一个重要因素，且其乳制品份额占据很大的市场［联合国粮农组织（FAO），2008］。奶牛、山羊和绵羊的乳产量占据全球乳产量的87%（FAO，2008）。然而，在一些国家，小的泌乳动物在其营养供给及经济发展中起着重要的作用[6]。但是，生物多样性以及未充分利用物种乳对营养健康的贡献一直被忽视，且不同畜种乳潜在的营养

价值还没有被充分地发掘出来。奶牛乳、水牛乳、山羊乳、绵羊乳和骆驼乳及其乳制品占据人类乳品消费的大部分，但从乳糖含量和乳蛋白结构的角度而言，驴乳的成分更接近于人乳[6-7]。水牛乳产量占世界乳产量的第二位，约占世界乳产量的 13%（FAO，2010）。而在中国的山区，蒙古、俄罗斯、尼泊尔、印度、不丹、塔吉克斯坦和乌兹别克斯坦等国，由于没有其他的牛科类可以饲养，故大部分的人类主要依靠牦牛来提供肉类和乳类[8]。此外，麋鹿和驼鹿等的乳也是一些偏远山区居民的乳品来源，而有关其成分等的相关信息却甚少[6]，且对其营养价值及对人类健康所起的作用也了解甚少。

FAO 公布（2008），截至 2011 年，全世界奶牛乳、山羊乳、绵羊乳、水牛乳和骆驼乳的产量分别是 6.1 亿吨、1 586万吨、926 万吨、9 302万吨和 226 万吨，其中奶牛乳、水牛乳和山羊乳产量分别占乳总产量的 84%、13%和 2%，如图 1-1所示。

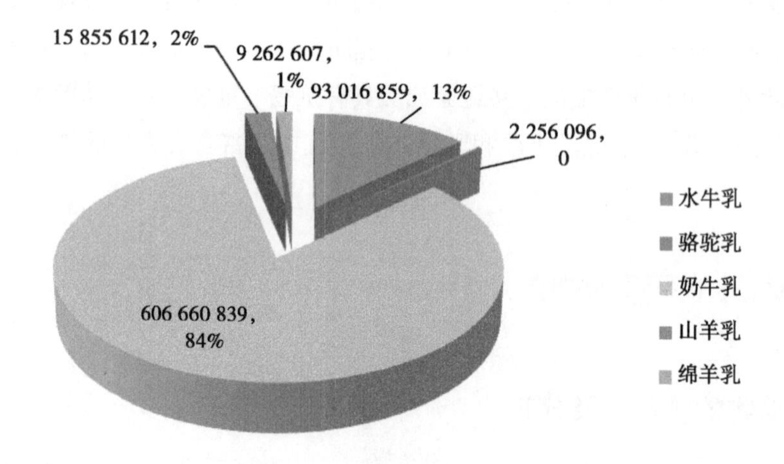

图 1-1　2011 年不同畜种乳产量（吨）及所占比例

Fig. 1-1　Milk yield and proportion of different species in 2011（1 000 kg）

1.1.2　不同畜种乳成分

1.1.2.1　常规乳成分

不同畜种乳成分如表 1-1 所示，不同畜种乳成分含量存在一定的差别。其中，马乳和驴乳中的乳蛋白、乳脂肪含量和能量值都较低，乳糖含量则较其他畜种乳高；而牦牛乳和绵羊乳中乳蛋白和乳脂肪含量均较其他畜种高。

不同畜种乳中矿物质元素含量见表 1-2，不同畜种乳矿物质元素含量差异显著，由于乳中的矿物质元素参与机体的很多代谢过程，尤其是钙（Ca）、钠（Na）、镁（Mg）、钾（K）、磷（P）、碘（I）等元素。而乳中元素的含量同时影响着乳的生物学价值。Gaucheron（2005）的研究指出，乳中的 Ca 与酪蛋白结合后极易被消化吸收[9]，而这与乳中酪蛋白的浓度有较强的相关性[10]。水牛乳和山羊乳中的 Ca 元素含量较高，而马乳中的 Ca 含量较少。

表 1-1　不同畜种乳成分

Table 1-1　Different species milk components

乳成分 Milk components	蛋白 Protein (g/100 g)	脂肪 Fat (g/100 g)	乳糖 Lactose (g/100 g)	能量 Energy (kJ/100 g)	灰分 Ash (g/100 g)
奶牛 Cow	3.2	3.3	5.1	317~373	0.7
水牛 Buffalo	4.0	7.4	4.4	345	0.8
牦牛 Yak	5.2	6.8	4.8	368	0.8
绵羊 Ewe	5.73	6.99	4.75	593	—
山羊 Goat	3.26	4.07	4.51	302	—
骆驼 Camel	3.26	3.80	4.30	328	0.8
马 Mare	2.0	1.6	6.6	184~205	0.4
驴 Donkey	1.6	0.7	6.4	208~245	0.4

注：引自 Barlowska 等（2011）和 Medhammar 等（2012）[6,11]

表 1-2 不同畜种乳中矿物质含量

Table 1-2 Mineral contents in different species milk

	奶牛 Cow	水牛 Buffalo	牦牛 Yak	山羊 Goat	骆驼 Camel	母马 Mare
钙 Ca (mg/100 g)	113	191	129	132~134	114~116	95
磷 P (mg/100 g)	84	185	106	97.7~121	87.4	58
钠 Na (mg/100 g)	43	47	29	41~59.4	59	16
钾 K (mg/100 g)	132	112	95	152~181	144~156	51
镁 Mg (mg/100 g)	10	12	10	15.8~16	10.5~12.3	7
铁 Fe (μg/100 g)	30	170	570	7~60	230~290	100
铜 Cu (μg/100 g)	30	20	410	5~80	240	50
锌 Zn (μg/100 g)	400	500	900	56~370	530~590	200

注：引自 Barlowska 等（2011）和 Medhammar 等（2012）[6,11]

乳中富含水溶性维生素和脂溶性维生素。Park（2007）的研究指出，山羊乳和绵羊乳中均含有丰富的维生素 A[12]。此外，Barlowska 等（2011）报道指出，山羊乳中维生素 A、泛酸、硫胺素、核黄素和烟酰胺含量丰富，而维生素 B_{12} 和叶酸的含量仅为奶牛乳的 1/6[11]，而这两种维生素的缺乏易引起人类贫血症的发生[12-13]。骆驼乳维生素 C 的含量很高，是奶牛乳中维生素 C 含量的 30 倍，是人乳中维生素 C 含量的 6 倍[11]。

此外，根据乳畜在系谱树上占据的位置，可以将不同的乳畜进行归类，如图 1-2 所示[14]。马和驴属于奇蹄目；牛科类动物、山羊和绵羊属于偶蹄目和反刍亚目，而牦牛在系谱树上与奶牛同属于牛属；骆驼虽具有反刍特性，却属于胼足亚目，而非反刍动物。Smiddy 等（2012）的研究表明，许多乳成分的含量及分布也呈物种种属内相似性以及种属间差异性，马乳和驴乳中的乳蛋白、乳脂肪和乳糖的含量相近，而与其他畜种乳中的差异显著；甘油三酯的组成上，马乳和驴

乳中的甘油三酯相似性较高，而奶牛乳、绵羊乳、水牛乳、山羊乳中的甘油三酯组成相似[14]；Blasi 等（2008）研究指出，水牛和奶牛乳脂中 n-3 脂肪酸的含量是 n-6 脂肪酸含量的 9.5~10.0 倍，山羊和绵羊乳脂中相对应的比例是 3.2 和 3.4，而驴乳中相对应的比例仅为 1.8[15]；相对于乳蛋白而言，采用二维凝胶测定的水牛和奶牛乳蛋白的图谱相似，但在骆驼乳蛋白中检测出许多 κ-酪蛋白（κ-CN）亚型，且没有检出 β-乳球蛋白（β-Lg），而马乳中未检出 κ-CN，山羊乳的酪蛋白则主要以 β-Lg 为主[16]。

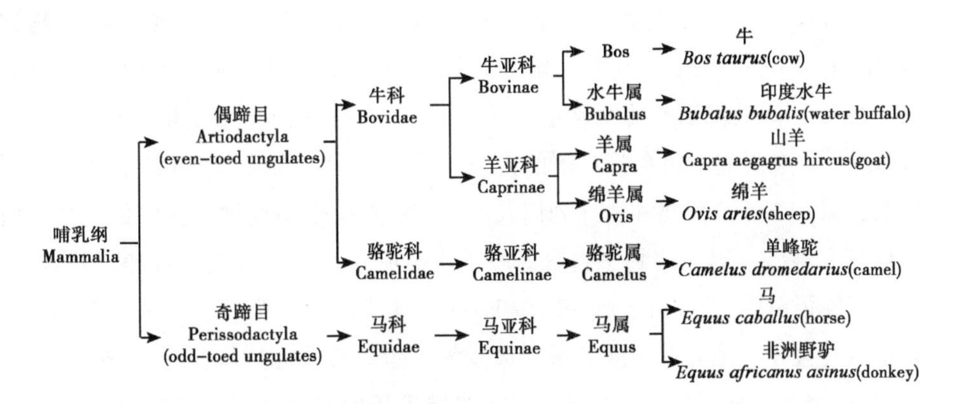

图 1-2 不同哺乳动物的系谱树（Barłowska 等，2011）

Fig. 1-2 Phylogenetic tree of the mammalian species（Barłowska et al，2011）

1.2 乳成分内在特征的研究

1.2.1 乳蛋白特征的研究

乳蛋白作为牛奶重要的营养指标，其含量及组成直接影响牛奶的品质。乳蛋白主要包括酪蛋白和乳清蛋白，其中酪蛋白含量约为总蛋白的 80%，主要以 α_{s1}-酪蛋白（α_{s1}-CN）、α_{s2}-酪蛋白（α_{s2}-CN）、β-酪蛋白（β-CN）和 κ-酪蛋白为主（κ-CN），而乳清蛋白则主要以 β-乳球蛋白（β-LG）、α-乳白蛋白（α-LA）、免疫球蛋白和乳铁蛋白等蛋白为主。其中 α_{s1}-CN、α_{s2}-CN、β-CN、κ-CN

存在于许多动物的乳中，并含有许多亚型[17]。

二维凝胶电泳（2-DE）是目前串联质谱蛋白质组学常用的一种分析技术，分析研究乳蛋白表达情况较为直接。D'auria 等（2005）通过 2-DE 研究证实其作为蛋白质学的一种检测技术可以有效地区别不同哺乳动物物种乳蛋白[18]。Aslam 等（1994）运用 2-DE 技术分析处于干乳期的奶牛乳蛋白表达模式得出，酪蛋白的含量在整个干乳期降低[19]。Smolenski 等（2009）通过对泌乳高峰期、泌乳初期及患有乳房炎的脱脂牛奶、乳清及乳脂球膜蛋白采用液相色谱-串联质谱法（LC-MS/MS）、2-DE 及基质辅助激光解吸电离飞行时间质谱法进行测定（MALDI-TOF-MS），LC-MS 共鉴定 2 903个肽类，通过 2-DE 鉴定出 2 770个蛋白点，95 个不同的基因产物被确定，其中 53 个通过 LC-MS/MS 和 57 个通过 2-DE 确定[20]。有研究通过免疫吸附的方法移除高丰度蛋白质（β-CN 和免疫球蛋白）后，采用 2-DE 分离和微序列以及质谱方法对奶牛初乳和常乳中低丰度蛋白质进行鉴定得出，奶牛初乳中含有血纤维蛋白原 β-链、α-抗胰蛋白酶和阿朴脂蛋白等特异蛋白质[21]。由于 α_{s1}-CN 和 β-CN 属于高丰度蛋白，在采用 2-DE 分离乳蛋白时会有一部分低丰度的蛋白被掩盖，如 κ-CN，但可通过利用 α_{s1}-CN 和 β-CN 缺少半胱氨酸的特点，采用 κ-CN 半胱氨酸的亲和富集的方法消除 α_{s1}-CN 和 β-CN 对其的影响，此外，经过 MALDI-TOF-MS 鉴定后，牛乳中含有 16 种 κ-CN 形式，其中在等电点为 4.47～5.81 含有 10 种 κ-CN 形式，且表现为 A 和 B 两种变异体，并通过串联质谱鉴定，得出其中含有 κ-CN 磷酸化形式，含有磷酸基团为 1～3 个[22]。Reinhardt 和 Lippolis（2008）对奶牛乳脂球膜（milk fat globule membrane，MFGM）蛋白运用 SDS-PAGE 分离，微毛细管高效液相色谱耦连串联质谱的方法鉴定，在鉴定的 120 个蛋白质中有 71%是膜联蛋白，仅有 15%的蛋白质是与鼠或人 MFGM 蛋白研究中鉴定的蛋白质相同。对鉴定的蛋白质进行归类发现：23%为膜-蛋白转运蛋白质，23%为细胞信号蛋白质，11%为脂肪转运-代谢蛋白质，9%为运输蛋白质，7%为蛋白质合成蛋白，4%为免疫蛋白和 21%未知功能蛋白。这些与细胞信号或膜-蛋白转运相关的蛋白质可为分析 MFGM 的分泌机制提供基础；功能免疫蛋白质 CD14、Toll 样受体（toll like receptor，TLR）TLR2 和 TLR4 可为监测乳腺组织的感染提供直接的证据[23]。为了进一步改善乳脂肪球膜蛋白的提取纯化，以获取更多的蛋白质表达信息，Vanderghem 等（2008）采用包括 CHAPS 在内的四种方法溶解抽提脂肪球膜蛋白，质谱鉴定了 2-DE 凝胶中 95 个蛋白点，涉及脂肪分泌和传递、蛋白信号转导和调节等功能[24]。采用标记蛋白质组学技术对初乳和常乳 MFGM 蛋白的研究发现，相对于初乳而言，常乳中有 26 个蛋白表达量增加，包括黏蛋白 1、黏蛋白 15 和嗜乳脂蛋白等；19 个蛋白的表达量降低，包括载脂蛋白 A1、载脂蛋白 E 等[23]。此后，Affolter 等（2010）将 MFGM 分为乳清和乳酪蛋白两种组分，经液相色谱串联质谱分别鉴定了 244 个和 133 个蛋白，并将这些蛋白据其功能分为信号转导、蛋白水解、免疫和防御及脂肪代谢相关蛋白[25]。

乳蛋白中酪蛋白的含量占 80%，酪蛋白含量及组分也是检测乳蛋白特征的重要指标。由于色谱技术的发展，近几年的研究中许多学者采用该项技术进行乳蛋白含量及特征的分析。Bramanti 等（2002）通过色谱分离技术，在 UV 检测器 280 nm 处酪蛋白含量（0.5～40 μmol/L）和峰面积线性相关，α-CN、β-CN 和 κ-CN 的检测限为 0.33～0.65 μmol/L，α_{s1}-CN、α_{s2}-CN、β-CN 和 κ-CN 的 RSD 值为 4.4%～6.2%[26]。Bonizzi 等（2009）通过反相高效液相色谱法从乳样中分离 α_{s1}-CN、α_{s2}-CN、β-CN 和 κ-CN，并通过电喷雾离子质谱技术进行定量分析，结果显示，该项技术可以高效的分离奶牛乳蛋白并进行量化[27]。此外，采用高效液相色谱技术测定其他物种乳蛋白的研究较多。Chen 等（2004）采用高效液相色谱串联电喷雾离子阱质谱的方法测定了奶牛乳清和山羊乳清中蛋白的含量，得出两者乳清蛋白总离子流图谱存在一定的差异，并指出在牛乳中掺入山羊乳时，最低检出限为 5%[28]。Feligini 等（2009）通过反相高效液相色谱串联质谱法测定水牛乳酪蛋白的试验得出，β-CN 和 α_{s2}-CN 均存在一个峰值，α_{s1}-CN 和 κ-CN 则存在多个峰值[29]。通过微芯片电泳及反相高效液相色谱法（reversed phase high-performance liquid chromatography，RP-HPLC）测定水牛乳中乳蛋白的结果显示，水牛乳中 β-CN（18266Da）、α-CN（14236Da）和血清白蛋白（66397Da）的含量分别是 4.04 g/L、2.45 g/L 和 0.35 g/L[30]。Bonfatti 等（2008）的研究结果显示，通过反相高效液相色谱法可有效的分离和量化牛乳中主要的蛋白组分及其基因型，鉴定出 β-CNI、α_{s1}-CNB 和 α_{s1}-CNC 等基因型的存在，纯化的蛋白基因型用于校准并均在 214 nm 处有不同的吸收峰[31]。此外，通过反相高效液相色谱法测定水牛乳蛋白的含量及基因型，分析结果显示，α_{s1}-CN、α_{s2}-CN、β-CN 和 κ-CN 占总酪蛋白的含量分别为 32.2%、15.8%、36.5% 和 15.5%，并指出 β-CN 的含量是 α-CN 含量的 1.3 倍，证实了地中海水牛乳中 κ-CN 存在多态性，并通过 RP-HPLC 测定出 α_{s1}-CN 及 κ-CN 的不同基因型[32]。

1.2.2 乳蛋白含量的研究

乳作为一种优质的食物来源，可以很好地满足新生哺乳类动物的营养需要[33]。乳蛋白主要成分的含量对其营养价值和加工工艺有重要的影响[34]。研究报道指出，所有的蛋白组分均具有不同的甚至特殊的生物学功能[35]。牛乳及水牛乳中主要酪蛋白和乳清蛋白及其比例均已有研究报道[32]。此外，有关羊乳及牦牛乳蛋白含量及组成的研究也有报道[36-38]。Zicarelli（2004）研究报道，水牛乳中酪蛋白含量为 4.0 g/100 g[39]，骆驼乳中的酪蛋白含量为 2.21 g/100 g[40]，而山羊乳中的酪蛋白含量为 2.81 g/100 g[41]。Guo 等（2007）的研究表明，人乳中乳清蛋白的含量为 0.69~0.83 g/100 g，马奶中乳清蛋白的含量为 0.74~0.91 g/100 g，牛乳中乳清蛋白的含量为 0.55~0.70 g/100 g，驴奶中乳清蛋白的含量为 0.49~0.80 g/100 g[42]；而绵羊乳中乳清蛋白和酪蛋白的含量均较高，分别为 1.02 g/100 g 和 4.18 g/100 g[43]。驴乳和马乳中酪蛋白的含量较低，分别为 0.64~1.03 g/100 g 和 0.94~1.2 g/100 g[42]。Zicarelli（2004）研究指出，奶牛乳和水牛乳中的 α-CN 含量较高，分别占乳中总酪蛋白的 38.4% 和 30.2%[39]。此外，Bobe 等（2007）研究报道，牛乳中酪蛋白的含量占总乳蛋白的 86%，其中 α_{s1}-CN 为 31.42%、β-CN 为 10.37%，而乳清蛋白为 14.12%[44]。

早期研究乳蛋白组成及含量的测定方法种类较多，包括免疫法（immunological）、电泳（electrophoretic）、毛细管电泳（capillary electrophoretic）和高效液相色谱法（high-performance liquid chromatography，HPLC）等[45-50]。此外，近年来随着蛋白质组学技术的发展，其也被用于蛋白的分离与鉴定，且蛋白质组学技术可以进行蛋白翻译后修饰的检测[51]，蛋白质组学技术也可以同时分离检测大量的乳蛋白[52]。Yang 等（2013）通过蛋白质组学技术同位素标记相对和绝对定量

(isobaric tag for relative and absolute quantification, iTRAQ) 分离鉴定了奶牛、牦牛、山羊和骆驼乳清蛋白，且共鉴定211个蛋白，其中基于基因本体论的诠释方法将113个蛋白进行功能分类，其中包括分子功能、细胞组成和生物学功能[53]。然而，蛋白质组学技术仍在发展且存在一定的局限性[51]。

然而，大量的研究表明，HPLC是乳蛋白分离量化的最可靠的方法[31-32,54-55]。Bonfatti等（2008）通过RP-HPLC分离鉴定了奶牛乳蛋白，且另有研究通过RP-HPLC定性且定量分析了水牛乳蛋白的多样性[32]。此外，Feligini等（2009）通过RP-HPLC-MS（RP-HPLC-mass spectrometry）鉴定分析得出，在水牛乳蛋白中存在α_{s1}-CN和κ-CN的异构体，但β-CN和α_{s2}-CN只存在一种构型[29]。Aschaffenburg和Sen（1963）研究表明，水牛乳蛋白中α_s-CN、β-CN和κ-CN的含量是相对的，且有别于奶牛乳[56]。有关山羊乳蛋白的研究也有报道，Trujillo等（2000）通过RP-HPLC分析鉴定了山羊乳蛋白及其基因型[57]。Li等（2010）通过RP-HPLC检测的结果显示，相比奶牛乳，牦牛乳中含有较高的乳蛋白，且酪蛋白组成含量也较高。然而，基于物种属性不同，上述畜种乳中蛋白组分及含量存在一定的差异[38]。且随着蛋白质组学方法在乳蛋白组成分析中的逐步应用，研究发现奶牛、马、水牛、绵羊、山羊和牦牛等不同畜种间乳蛋白质谱图谱存在规律性差异[28,58-59]。

1.2.3 乳脂球膜蛋白的研究

牛乳成分复杂，含有蛋白、脂肪、碳水化合物、维生素和矿物质。乳脂是牛乳的重要成分，是由乳腺上皮细胞内质网合成的以甘油三酯为中心外包有复杂的脂质双分子层膜。当脂滴释放进入牛乳时，乳脂球中的不同细胞质成分保留到不同的膜层中[60]。因此，乳脂主要由中性脂质、胆固醇、极性脂质和蛋白混合物组成[61]。尽管MFGM蛋白仅占牛乳总蛋白的1%~4%，但与其他牛乳成分相比具有更复杂的多样性[62]。由于乳脂球的功能与营养特性，许多研究对它们的蛋白成分进行了分析[63-64]。首先，MFGM在牛乳水相脂质中起到了乳化剂的作用；其次，一些MFGM蛋白具有广泛的生物学功能，如阻止病原体附着、参与抗菌防御活动[65-67]。因此，MFGM成分的研究逐渐引起广泛关注。

对于MFGM蛋白组的研究主要集中于人乳和牛乳。目前，采用蛋白质组学的方法已对人乳MFGM中大量蛋白进行了成功鉴定[68-70]。牛乳MFGM蛋白组成的研究最初是通过一维SDS凝胶电泳结合液相色谱或串联质谱进行鉴定[71]。随后，一些程序被应用于提取和溶解疏水性MFGM蛋白，这些蛋白通过2-DE结合质谱技术得以分离鉴定[24,72]。此外，采用蛋白质组学技术对初乳和常乳[23]、脱脂乳和奶油[73]、富含MFGM的乳清蛋白浓缩物和脱脂乳蛋白浓缩物[25]、金黄色葡萄球菌感染乳房炎牛乳和健康牛乳中的MFGM蛋白组差异性进行了鉴定分析[74]。此外，定量分析包括无标记和二甲基标记质谱技术也已用于牛乳MFGM蛋白的研究[25,61,75]。这些研究极大地促进了牛乳MFGM蛋白组的研究分析。

小型泌乳动物（如牦牛和骆驼）所产的乳汁在特定地区具有重要的营养价值和经济价值[6]。除此之外，马、骆驼和牦牛乳可以作为人类饮食中的功能性食品[76]。因此，已有报道对山羊[77-78]、水牛[79]、绵羊[80]、马[81-82]和骆驼[83]这些小型泌乳动物的MFGM蛋白组学图谱进行了研究。此外，有研究采用2-DE结合质谱技术对两种牛科动物（契安尼那牛和荷斯坦牛）的MFGM蛋白进行了比较分析[84-85]。除此之外，已有研究采用SDS-PAGE技术对山羊、绵羊、马和骆驼等畜种的主要MFGM蛋白进行了比较分析[86]。山羊、牛和人乳MFGM中的主要蛋白通过2-DE结合质谱技术进行了鉴定[87]。尽管先前已存在对MFGM蛋白的研究，但非牛源性MFGM蛋白的研究相对较少且不全面。因此，对MFGM蛋白组成和功能的全面了解受到了一定的制约。但是，由于对于小型泌乳动物（尤其是驴、牦牛和骆驼）乳需求的不断增加[76]，为了更好地了解MFGM蛋白的生物学功能和潜在的营养特性，需要对小型泌乳动物MFGM蛋白进行进一步的深入分析研究，以揭示不同物种MFGM蛋白的差异性，为评定乳品质提供依据。

1.2.4 乳脂肪的研究

乳脂肪主要是由一个分子的甘油和三个分子脂肪酸组成，其中的乳脂球平均直径为0.1~18 μm[88]。Barlowska等（2011）的研究指出，乳脂肪球的直径与物种种属有关。其中，水牛乳的脂肪球直径为8.7 μm，是所有泌乳动物脂肪球直径最大的；而骆驼乳和山羊乳脂肪球的直径最小，分别为2.99 μm和3.19

μm[11]。其他的泌乳动物，如牛乳中脂肪球直径是0.92~15.75 μm，平均直径为3.51 μm，其中90%的乳脂球直径小于6.42 μm；山羊乳的脂肪球的直径范围是0.73~8.58 μm，平均直径2.76 μm，其中90%乳脂球直径小于5.21 μm[89]。此外，有关乳脂肪酸构成的研究报道也有很多，Jensen等（2002）的研究得出，牛乳中60%~70%的是含碳原子数为14、16和18的饱和脂肪酸，而主要以油酸为主的单不饱和脂肪酸约为25%~30%，其余大约4%是多不饱和脂肪酸，主要为亚麻酸和亚油酸[90]。“理想”牛奶中饱和脂肪酸含量应不超过总脂肪酸的8%，多不饱和脂肪酸应在10%，而其余脂肪酸为单不饱和脂肪酸（82%），这与典型乳脂的组成有很大的差异。而乳中的甘油三酯约占乳总脂肪的98%，不同畜种乳的甘油三酯成分含量不同（表1-3）[14]，且相应的结果显示，骆驼乳脂中主要的甘油三酯的组分为C48、C50和C52，而水牛乳、奶牛乳、山羊乳和绵羊乳中甘油三酯组分含量较多的为C32和C40，马乳和驴乳甘油三酯组分含量较高的为C44[14]。

表1-3 不同畜种乳中甘油三酯的含量（%）

Table 1-3 Triglyceride composition content in different species milk（%）

	奶牛	山羊	绵羊	骆驼	马	驴	水牛
C24	0.06 ± 0.03^{c}	0.08 ± 0.02^{b}	0.08 ± 0.03^{b}	0.10 ± 0.05^{a}	0.08 ± 0.03^{bc}	0.09 ± 0.06^{ab}	0.08 ± 0.04^{bc}
Chol	0.42 ± 0.04^{a}	0.54 ± 0.06^{bc}	0.47 ± 0.10^{cd}	0.59 ± 0.25^{b}	0.88 ± 0.35^{a}	0.82 ± 0.33^{a}	0.47 ± 0.10^{cd}
C26	0.24 ± 0.02^{c}	0.32 ± 0.04^{b}	0.38 ± 0.15^{a}	0.02 ± 0.01^{d}	0.24 ± 0.11^{c}	0.25 ± 0.14^{c}	0.40 ± 0.12^{a}
C28	0.56 ± 0.05^{c}	0.78 ± 0.11^{ab}	0.86 ± 0.31^{a}	0.01 ± 0.00^{d}	0.51 ± 0.28^{c}	0.57 ± 0.32^{c}	0.75 ± 0.24^{b}
C30	1.12 ± 0.13^{c}	1.69 ± 0.14^{a}	1.57 ± 0.53^{a}	0.03 ± 0.01^{d}	1.04 ± 0.55^{c}	1.18 ± 0.54^{bc}	1.30 ± 0.40^{b}
C32	2.46 ± 0.31^{c}	3.02 ± 0.29^{a}	2.65 ± 0.84^{bc}	0.09 ± 0.03^{a}	1.38 ± 0.70^{d}	1.53 ± 0.68^{d}	2.82 ± 0.82^{ab}
C34	5.72 ± 0.59^{b}	5.43 ± 0.36^{b}	4.85 ± 1.03^{c}	0.24 ± 0.05^{a}	2.23 ± 0.97^{d}	2.08 ± 0.72^{d}	6.99 ± 1.81^{a}
C36	10.44 ± 0.60^{b}	9.16 ± 0.56^{c}	8.33 ± 1.06^{d}	0.46 ± 0.09^{a}	3.42 ± 1.04^{a}	3.33 ± 0.84^{e}	12.47 ± 2.35^{a}
C38	12.27 ± 0.46^{c}	11.85 ± 1.17^{bc}	11.88 ± 1.08^{c}	0.64 ± 0.15^{f}	4.40 ± 1.30^{d}	4.44 ± 1.15^{d}	13.55 ± 1.39^{a}
C40	9.66 ± 0.52^{c}	12.14 ± 1.94^{a}	11.08 ± 1.04^{b}	0.78 ± 0.61^{f}	5.76 ± 1.71^{d}	5.16 ± 1.31^{e}	9.49 ± 1.25^{c}
C42	6.83 ± 0.62^{d}	11.14 ± 0.47^{a}	7.97 ± 1.38^{c}	1.80 ± 1.29^{f}	8.86 ± 2.02^{b}	7.63 ± 1.19^{c}	5.77 ± 0.86^{a}
C44	6.55 ± 0.73^{a}	9.86 ± 0.92^{c}	7.17 ± 1.26^{d}	5.20 ± 2.12^{a}	11.49 ± 1.91^{a}	10.55 ± 1.43^{b}	5.40 ± 1.01^{f}
C46	7.43 ± 0.74^{d}	7.78 ± 0.88^{d}	6.81 ± 0.82^{e}	12.48 ± 2.22^{a}	10.41 ± 1.26^{b}	9.00 ± 1.45^{c}	6.68 ± 1.21^{e}

（续表）

	奶牛	山羊	绵羊	骆驼	马	驴	水牛
C48	9.26±0.56[c]	7.08±0.68[d]	7.38±0.70[d]	21.60±1.91[a]	10.02±1.83[b]	7.10±0.48[d]	8.79±1.28[c]
C50	11.33±0.59[cd]	8.22±0.96[f]	9.66±1.83[e]	25.79±2.25[a]	13.19±3.19[b]	11.43±3.07[c]	10.52±1.90[de]
C52	10.33±2.06[cd]	7.26±1.14[e]	11.16±3.02[c]	20.37±2.84[a]	17.39±5.50[b]	21.01±5.47[a]	9.39±4.13[de]
C54	4.90±1.40[d]	3.27±0.07[a]	7.18±2.58[e]	9.07±1.90[b]	8.70±3.99[b]	13.32±2.07[a]	4.64±3.05[d]

注：引自 Barłowska 等（2011）；

a-f 肩标表示差异显著（$P<0.01$）

此外，驴乳的脂肪酸中中链饱和脂肪酸 C8:0、C10:0 和 C12:0 的含量较高，而 C14:0 和 C16:0 含量较低，但 C18:2 和 C18:3 含量均高于反刍动物的乳。Baelowska 等（2011）的研究指出，绵羊乳的 C4:0 含量及共轭亚油酸的含量均高于奶牛乳和山羊乳[11]。

大量的试验证实，气相色谱法（gas chromatograph，GC）、HPLC 技术是检测乳脂肪酸最常用的检测技术，且证实乳脂肪的变异程度较乳蛋白大，主要受到奶牛日粮、泌乳阶段、品种等因素的影响，但 Iverson 等（1989）的研究得出，各物种乳中脂肪酸 C12:0/C10:0 的比值较恒定，其中奶牛 C12:0/C10:0 的比率约为 1.16，而山羊乳的比率为 0.46，绵羊乳的比率为 0.58，据此可区分奶牛与山羊、绵羊乳的乳脂肪[91]。因此，脂肪酸的组成及其含量是衡量和鉴别乳脂肪质量的潜在标志分子。此外，由于甘油三酯对人体的一些生理功能，其结果的研究也成为当前研究的一个重点。HPLC 是检测乳脂中甘油三酯含量的主要检测技术，而薄层层析法（thin-layer chromatography，TLC）则是研究其在甘油结合脂肪酸酯化后，所结合脂肪酸的结合位点的主要检测技术[91]。Haddad 等（2010）通过采用 TLC 对骆驼乳脂肪甘油三酯的测定得出，碳原子数为 10~16 个的脂肪酸在酯化为 TAG 时主要分布结合在甘油的第二个结合位点，而碳原子数超过 16 个的脂肪酸在酯化为 TAG 时主要分布结合在 1、3 这两个结合位点[92]。但 Blasi 等（2008）采用 TLC 对驴乳脂中甘油三酯的测定得出，饱和脂肪酸（saturated fatty acid，SFA）在酯化为 TAG 时主要结合在甘油的第 3 个结合位点，而单不饱和脂肪酸（monounsaturated fatty acid，MUFA）则主要结合在第 2 个结合位点。

这可能与不同畜种有关，为区分不同畜种及建立荷斯坦奶牛特定的甘油三酯脂肪酸构成奠定基础[15]。

有关奇数碳链支链脂肪酸（odd and branched chain fatty acids，OBCFA）的报道，在很多不同的领域都有相关的研究。一些研究指出，支链脂肪酸（branched chain fatty acids，BCFA）具有抗癌作用[93-94]，是由于这些癌细胞的生物合成相比健康的细胞更依赖于脂肪酸[95]，且鉴于此，一些学者进行了一些 OBCFA 相关的抗癌治疗方面的研究[96]。Vlaeminck 等（2006）报道，BCFA 中的 *iso*-C15:0 在体外和体内试验中均可很好的抑制癌细胞的生长[94]。此外，也有研究报道 *anteiso*-C15:0 也具有与 *iso*-C15:0 一样的抑制癌细胞生长的功能[93]。但是，当碳链增加后，即从 *iso*-16:0 以后的这些脂肪酸（Fatty acid，FA）的抗癌作用就会下降[97]。此外，乳脂肪中的另外一部分微量组成，如乳脂肪中的共轭亚油酸亦被作为一种潜在的抗癌因子而得到很大的关注[5]。由于反刍动物畜产品中的 OBCFA 的低熔点性而备受研究者的关注[98]。

在大多数的植物中，OBCFA 组成的含量一般都是微量级别的[99]。然而，在一些动物乳中和组织中，OBCFA 的含量和组成是不同的，如奶牛[100]、绵羊[101]、山羊[102]和海狸[103]。Keeney 等（1962）研究指出，乳中的 OBCFA 大部分是来自瘤胃微生物[104]，且近期的研究亦证明该结论[94,105]。当日粮中的 OBCFA 全部用于转化为乳中 OBCFA 时，只有 100 g/kg 的乳 OBCFA 来自日粮[106]。奶牛乳中 OBCFA 主要包含 *iso*-C13:0（tridecanoic acid isomers），*iso*-C14:0（tetradecanoic acid），C15:0，*iso*-C15:0，*anteiso*-C15:0（pentadecanoic acid），*iso*-C16:0（hexadecanoic acid），C17:0，*iso*-C17:0 和 *anteiso*-C17:0（heptadecanoic acid）[107]。山羊乳中的 BCFA 的检测亦有报道，且对奶牛和山羊乳中 BCFA 的差异性亦进行了检测[108]。此外，DePooter 等（1981）研究指出，尽管奶牛与山羊乳中具有奇数或偶数碳原子的顺式脂肪酸的数量相似，但其含量仍存在较大的差异[109]。反刍动物和单胃动物乳中的脂肪及其组成存在较大的差异，Devle 等（2012）的研究指出，在反刍动物乳中，绵羊乳中的 OBCFA 含量最多，占总 FA 的 5.5%，且在反刍动物与单胃动物乳脂肪中，占据主导地位的 OBCFA 组分均为 C15:0 和 C17:0[110]。有关牦牛乳[111-113]、娟姗牛乳[101,114-115]、马乳[116]和骆驼乳[117-118]中 FA 的研究分别均有报道。然而，有关牦牛乳、骆驼乳、马乳、水牛乳和娟姗牛乳中 OBCFA 的相关数据报道较有限。此外，随着对非牛科类哺乳动物乳需求量的不断增长[76]，有关不同畜种乳中脂肪组成和功能，其中包括 OBCFA 等的相关研究有待于进一步深入。

1.2.5 乳中矿物质元素的研究

有关乳中矿物质元素的研究报道较多，鉴于乳中元素除了对人体营养方面起一定的生物学功能，其中的某些元素对机体也有一定的毒副作用。因此，研究乳中元素的含量及组成是非常有必要的[119]。然而，乳中常量及微量元素的含量变异性较大，主要影响因素较多，其中与物种本身的基因遗传性、乳腺的分泌、健康状况及泌乳阶段均有关系[120]。

尽管奶牛、山羊和绵羊乳产量占全世界总乳产量的 87%（FAO，2008），但目前对非反刍动物乳的需求量也在增加，如对马乳的需求[76]。且在一些国家里，小型的泌乳动物在满足营养需要和促进当地经济发展中也起着重要作用[6]。然而，鉴于物种遗传性及生物多样性的原因，以物种水平定义食品的研究需要进一步深入[121-122]。随着检测技术的不断发展，检测的灵敏度及准确度也在不断的提高。截至目前，已有较多的关于奶牛乳中元素测定的研究报道，且亦有一些研究采用的是电感耦合等离子体质谱（inductively coupled plasma mass spectrometry，ICP-MS）技术[119-120,123-127]。Khan 等（2006）通过分光光度计测定结果显示，山羊和绵羊乳中的元素 Ca、K、Mg、Na 和 Zn 含量存在显著的差异性[128]。Güler（2007）通过采用 ICP-MS 测定了山羊乳中 24 种元素的含量[129]。但有关骆驼和马乳中元素含量的报道较少，且没有同时测定比较分析奶牛乳、水牛乳、牦牛乳、娟姗牛乳、骆驼乳和马乳中元素含量的研究报道。此外，ICP-MS 检测技术作为具有高敏感度和高选择性的可同时分析多种元素的一种快速的标准检测技术已运用于很多研究领域[33,130-133]。因此，采用 ICP-MS 技术快速、准确的检测不同畜种乳中的元素含量，可为进一步评定畜种乳营养价值提供一定的基础信息。

1.2.6 乳中 microRNA 的研究

过去十年内，microRNA 越来越明确的被证明是小的非编码 RNA 的一大类，其通过调节基因的表达参与广泛的细胞活动过程。在 2000 年前，有关 microRNA 的研究都仅限于非哺乳或非脊椎动物。经过约 10 年的研究，人类已经发现哺乳动物中存在 microRNA，并探索了它们对疾病治疗方面的作用。

2000 年，let-7 的发现，开启了人类关于 microRNA 的研究，证明了 microRNA 在许多物种中进化保守并广泛表达[134-136]；2007 年，出现了 microRNA 重要性的第一个证据，Dicer 酶的纯合性缺失破坏了小鼠胚胎的发育[137]；2009 年，在人和动物的血清和血浆中检测 microRNA 开启了应用 microRNA 诊断各种疾病的可能性[138]（图 1-3）。

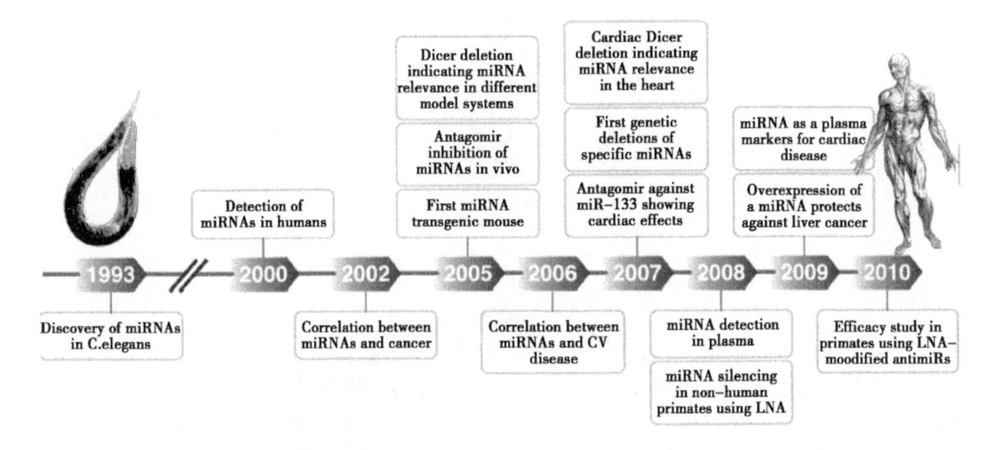

图 1-3 microRNA 研究的突破性进展（Eva van Rooij，2011）

Fig. 1-3 Breakthrough discoveries in miRNA biology（Eva van Rooij，2011）

microRNA 的生物合成受许多关卡的监管，初级 microRNA 的转录为初体 microRNA，称为 pri-microRNA，无论是从 annotated transcripts（例如，蛋白质编码基因的内含子，非编码基因的外显子，或者是非编码基因的内含子）还是基因组内的间隔区都能衍生出前体 microRNA，并可以编码单个或多个 microRNA[139-140]。初体 microRNA 折叠成发卡结构，其中包含不完全配对的碱基的茎和被加工成 60-nt到 100-nt 的发卡，这就是 pre-microRNA 即前体 microRNA[141-143]。pre-microRNA 从细胞核被 exportin5 转运到细胞质[144]，然后被核酸内切酶 Dicer 裂解成不完善的 microRNA-microRNA* 双链[145]，microRNA 链会变成一条成熟的 microR-

NA，而 microRNA* 链会被降解。成熟的 microRNA 会结合到由 RNA 诱导的沉默复合物（RISC）上，确定具体的靶基因并会诱导基因转录后沉默[146]（图 1-4）。

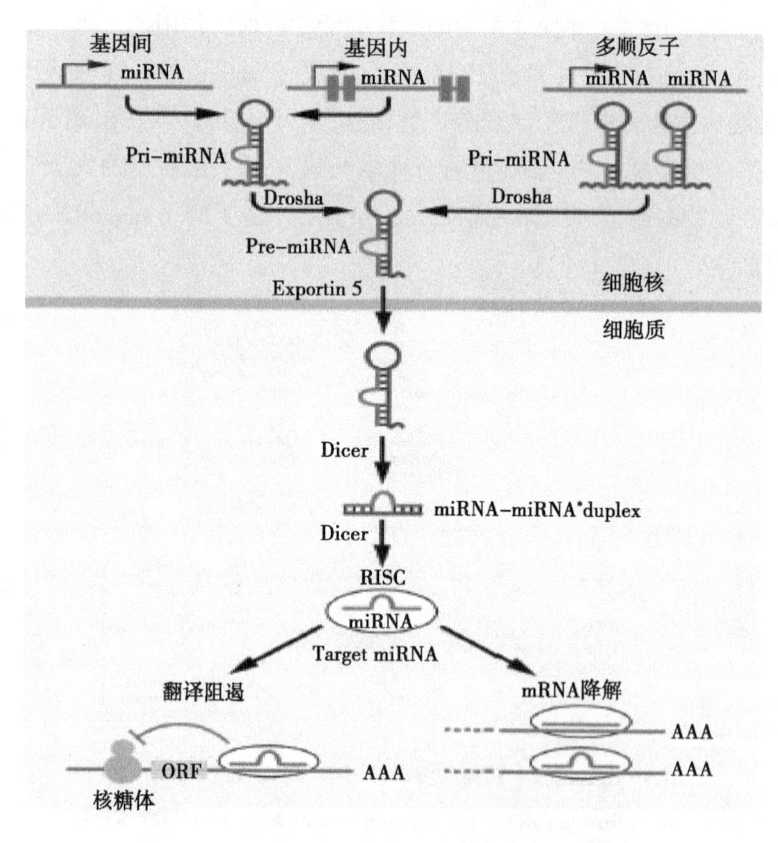

图 1-4　microRNA 的生物合成（Jensen，1995）

Fig. 1-4　miRNA biogenesis（Jensen，1995）

乳是哺乳动物新生儿的唯一营养来源，但是为了适合子代发育，不同的哺乳动物乳成分也不尽相同[147-148]。已有研究表明人类母乳中含有免疫相关的 microRNA[149]，在该研究中，器官特异性的 microRNA（血细胞、肌肉、胰腺和肝脏特异性）没有被检测到。microRNA 是内生的，非编码 microRNA 分子具有 19~24 个核苷酸的长度，通过与靶基因碱基互补配对在基因表达方面起着重要的调控作用，从而阻断了翻译或触发了靶基因的降解。大多数 microRNA 的生物学功

能是未知的，但估测大约30%以上的蛋白质编码基因受microRNA的调控[150]。

曾经有人报道过人乳中存在microRNA，最近其他研究者还报道了牛乳中也存在microRNA。Izumi等（2012）使用基因芯片和定量PCR方法研究了牛乳的microRNA，验证初乳和常乳之间的差异，初乳乳清中microRNA的浓度高于常乳乳清microRNA浓度。牛奶中共检测出102种microRNA，与免疫和发育相关的microRNA包括miR-15b、miR-27b、miR-34a、miR-106b、miR-130a、miR-155和miR-223。使用定量PCR检测牛乳中的这些microRNA，以上每种microRNA在初乳中的表达都显著高于常乳中的表达，证实了牛乳中microRNA的存在；此外，原料奶乳清中合成的microRNA易被降解，天然存在的microRNA和mRNA可以抵抗酸性条件和RNA酶的作用，也就是说，牛奶中的RNA分子是非常稳定的[151]。然而，乳中的microRNA是否具有一定的物种特异性，进而发挥一定的生物学功能，需要进一步深入研究。

2 研究内容、目的及意义

2.1 研究内容与技术路线

本书以不同畜种乳为研究对象，通过不同的组学技术和质谱检测技术测定其中的成分，为进一步了解不同畜种乳成分组成及含量提供一定的理论基础，并为区分不同畜种乳，甚至是掺假乳提供一定的依据。首先，通过 RP-HPLC 检测奶牛乳、水牛乳、牦牛乳和山羊乳中主要的乳蛋白组分和含量，并在本试验条件下建立不同畜种乳蛋白 RP-HPLC 图谱，为进一步通过不同畜种乳蛋白进行乳源的区分提供理论依据；其次，通过气相色谱—质谱联用（Gas Chromatography-Mass Spectrometer，GC-MS）技术检测奶牛乳、水牛乳、牦牛乳、娟姗牛乳、山羊乳、骆驼乳和马乳中的奇数碳链支链脂肪酸（OBCFA），从而得出不同种属乳脂中特征性 OBCFA 组分，为进一步开发 OBCFA 的功能及更好地区分不同物种乳提供依据；再次，通过 ICP-MS 分析测定了奶牛乳、水牛乳、牦牛乳、娟姗牛乳、山羊乳、骆驼乳和马乳中的 9 种元素（其中 4 种常量元素：K、Ca、Na 和 Mg，5 种微量元素：Mn、Co、Zn、Fe 和 Se）含量，为分析不同畜种乳常规成分含量提供数据信息；另外，通过 Label-Free 定量蛋白质组学方法分析奶牛乳、水牛乳、牦牛乳、娟姗牛乳、山羊乳和骆驼乳中的 MFGM 蛋白，为进一步了解不同畜种 MF-GM 蛋白的生物学功能和潜在的营养特性提供依据，并在本试验条件下建立不同畜种 MFGM 蛋白谱图；最后，通过转录组学方法分析新鲜奶牛与新鲜山羊乳清中特征性 microRNA 及反复冻融奶牛乳清与新鲜奶牛乳清 microRNA 的差异性，为进一步确定奶牛和山羊乳清中各自特征性 microRNA 及分析反复冻融对乳清中 microRNA 种类及表达量的影响，为进一步区分奶牛乳和山羊乳提供理论依据。

技术路线见图 2-1。

本研究共分为 5 个试验：

（1）基于 RP-HPLC 技术分析不同乳畜乳中主要乳蛋白组分。

（2）基于 GC-MS 技术分析不同乳畜乳中 OBCFA 及其组分含量。

（3）基于 ICP-MS 方法分析不同畜种乳中常量及微量元素含量。

（4）基于 Label-Free 定量蛋白质组学方法分析不同畜种乳 MFGM 蛋白。

（5）基于转录组学方法分析新鲜奶牛与新鲜山羊乳清中特征性 microRNA 及反复冻融奶牛乳清与新鲜奶牛乳清样 microRNA 的差异性。

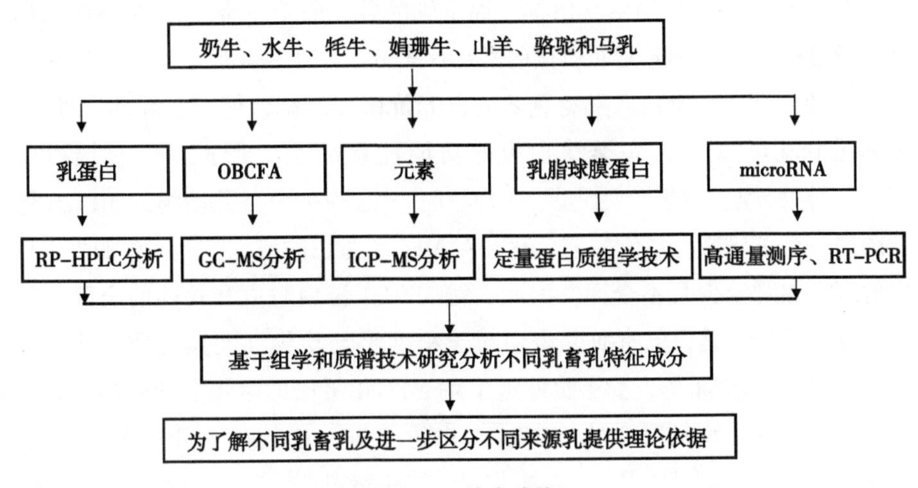

图 2-1　技术路线

Fig. 2-1　Technical route

2.2　研究的目的及意义

在一些国家，小的泌乳动物在其营养供给及经济发展中起着重要的作用。然而，生物多样性以及未充分利用物种对营养健康的贡献一直被忽视，且不同畜种乳的潜在的营养价值还没有被充分的发掘出来。水牛乳产量占据世界乳产量的第二位，约占世界乳产量的 13%（FAO，2008）。而在中国的山区和蒙古、俄罗斯、

尼泊尔、印度、不丹、塔吉克斯坦和乌兹别克斯坦等国，由于没有其他的牛科类可以饲养，故大部分的人类主要依靠牦牛来提供肉类和乳类[8]。此外，麋鹿和驼鹿等的乳也是一些偏远山区居民的乳品来源，而有关其成分等的相关信息却甚少[6]，且对其营养价值及对人类健康所起的作用也了解甚少。

随着蛋白质组学方法在乳蛋白组成分析中的逐步应用，研究发现奶牛、马、水牛、绵羊、山羊和牦牛等不同畜种间乳蛋白质谱图谱存在规律性差异[28,58-59]。Iverson 等（1989）的研究得出各畜种乳中脂肪酸 C12:0/C10:0 的比值较恒定，其中奶牛 C12:0/C10:0 的比率约为 1.16，而山羊乳的比率为 0.46，绵羊乳的比率为 0.58，据此可区分奶牛与山羊、绵羊乳的乳脂肪[91]。因此，脂肪酸的组成及其含量是衡量和鉴别乳脂肪质量的潜在标志分子。小型泌乳动物（如牦牛和骆驼）所产的乳汁在特定地区具有重要的营养价值和经济价值[6]。除此之外，马、骆驼和牦牛乳可以作为人类饮食中的功能性食品[76]。因此，已有报道对山羊[77-78]、水牛[79]、绵羊[80]、马[81-82]和骆驼[83]这些小型泌乳动物的 MFGM 蛋白组学图谱进行了研究。尽管先前已存在对 MFGM 蛋白的研究，但非牛源性 MFGM 蛋白的研究相对较少且不全面。因此，对 MFGM 蛋白组成和功能的全面了解受到了一定制约。但是，已有研究表明对于小型泌乳动物（尤其是驴、牦牛和骆驼）乳需求不断增加[76]。为了更好地了解 MFGM 蛋白的生物学功能和潜在的营养特性，需要对小型泌乳动物 MFGM 蛋白进行进一步的深入分析研究。此外，有大量研究表明人乳[152-154]、牛乳[151,155-156]和猪乳[157-158]中均存在 microRNA。Weber 等（2010）检测了 12 种人类体液中的 microRNA，其中包括人乳，但相比 Kosaka 等（2010b）的研究结果发现，不同的乳源背景，其所含有的 microRNA 种类和表达丰度不同[152]。因此，不同物种乳中的 microRNA 是否具有一定的物种特异性，需要进一步深入研究。

鉴于此，本研究的目的是通过研究不同畜种乳特征乳成分，揭示不同畜种乳特征差异性，为进一步阐释不同畜种乳营养成分及潜在的生物学功能的差异性，及评定乳品质提供一定的理论依据和基础信息，并为进一步区分不同畜种乳提供理论依据。

3 基于RP-HPLC技术分析不同乳畜乳中主要乳蛋白组分

3.1 引言

在早期研究乳蛋白组成及含量的测定方法种类较多，其中包括免疫法、电泳、毛细管电泳和 HPLC 等[45-50]。此外，近年来，随着蛋白质组学技术的发展，其也被用于蛋白的分离与鉴定，且蛋白质组学技术可以进行蛋白翻译后修饰的检测[51]，蛋白质组学技术可以同时分离检测大量的乳蛋白[52]。Yang 等（2013）通过蛋白质组学技术 iTRAQ 分离鉴定了奶牛乳、牦牛乳、山羊乳和骆驼乳清蛋白，共鉴定 211 个蛋白，其中基于基因本体论的诠释方法将 113 个蛋白进行功能分类，包括分子功能、细胞组成和生物学功能[53]。然而，蛋白质组学技术仍在发展且存在一定的局限性[51]。

大量的研究表明，HPLC 是乳蛋白分离量化的最可靠的方法[31-32,54-55]。Bonfatti 等（2008）通过反向高 RP-HPLC 分离鉴定了奶牛乳蛋白，且另有研究通过 RP-HPLC 定性定量分析了水牛乳蛋白的多样性[32]。此外，Feligini 等（2009）通过 RP-HPLC-MS 鉴定分析得出，在水牛乳蛋白中存在 α_{s1}-CN 和 κ-CN 的异构体，但 β-CN 和 α_{s2}-CN 只存在一种构型[29]。Aschaffenburg 和 Sen（1963）研究表明，水牛乳蛋白中 α_s-CN，β-CN 和 κ-CN 的含量是相对的，且有别于奶牛[56]。有关山羊乳蛋白的研究也有报道，Trujillo 等（2000）通过 RP-HPLC 分析鉴定了山羊乳蛋白及其基因型[57]。Li 等（2010）通过 RP-HPLC 的研究指出，相比奶牛乳，牦牛乳中含有较高的乳蛋白且酪蛋白组成含量也较高。然而，基于物种属性不同，上述物种乳中蛋白组分及含量存在一定的差异[38]。

鉴于 RP-HPLC 快速和自动化的分析、良好的分离效果、较高的分辨率、

较高的准确性及较高的重复性，本试验参照 Bonfatti 等（2008）[31] 的测定方法并进行一定的改进来分离和鉴定奶牛乳、山羊乳、水牛乳及牦牛乳蛋白，并检测其各自特有的色谱图谱，为进一步区分不同畜种乳甚至掺假乳提供一定的理论基础。

3.2 试验材料与方法

3.2.1 样品采集

于 2013 年 2 月至 4 月，从北京、河北、云南和青海养殖场分别采集奶牛乳样、山羊乳样、水牛乳样和牦牛乳样各 20 份，乳样于-20℃条件下保存，并利用装有干冰的泡沫箱运送至实验室，于-80℃条件下保存至分析。所有供体试验动物均处于健康状态，均处于泌乳日龄约为 4 个月。

3.2.2 试剂与蛋白标准品

盐酸胍（Guanidine hydrochloride，GdnHCl）（纯度>99%），二硫苏糖醇（纯度>99%），Bis Tris（纯度>98%）均购自 Sigma-Aldrich（St. Louis，MO，USA）；蛋白标准品 α-CN（纯度>70%），β-CN（纯度>90%），κ-CN（纯度>80%），α-La（纯度 85%），β-LgB（纯度>90%）和 β-LgA（纯度>90%）均购自 Sigma-Aldrich（St. Louis，MO，USA）；超纯水采自实验室（Milli-Q Plus System，Millipore，0.22 μm）。

3.2.3 样品的前处理

吸取乳样 0.8 mL，按 1：1 体积加入 pH 值=7 的 A 溶液（由 5.37 mM 的柠檬酸钠，19.5 mM 的二硫苏糖醇，6 M 的盐酸胍和 0.1 M 的 BisTris 混合而成），涡旋 10 s，室温培养 1 h 后，在 4℃条件下以 20 000 g 离心 15 min（Sigma，3~18 K），去除上层脂肪，吸取一定体积的下清液，加入 3 倍体积的 B 溶液（4.5 M 的盐酸胍乙腈溶液，其中乙腈：水：三氟乙酸=100：900：1），再通过 0.22 μm 滤

膜过滤，然后上机测定。

3.2.4 仪器与色谱条件

高效液相色谱仪为 Waters 2695（Waters Technologies，USA），安装有紫外检测器（Waters 2487），色谱柱为 PLRP-S（100 Å，5 μm column，150 mm×4.4 mm）。整个检测过程采用两种溶液进行梯度洗脱，A 溶液和 B 溶液，其中 A 溶液为含 0.1%三氟乙酸的水溶液，B 溶液为含 0.1%三氟乙酸的乙腈溶液，色谱柱温为 45℃，检测波长为 214 nm。梯度洗脱程序为 0→1.0 min，0→35% B 溶液；1.0 min→8.0 min，保持 35% B 溶液；8.0 min→16.0 min，35%→38% B 溶液；16.0 min→22.0 min，38%→42% B 溶液；22.0 min→22.5 min，42%→46% B 溶液；22.5 min→23.5 min，46%→100% B 溶液；23.5 min→30.0 min，100%→35% B 溶液，整个检测时间为 30 min。

3.2.5 标线的制定

将标准品 α-CN（纯度>70%），β-CN（纯度>90%），κ-CN（纯度>80%），α-La（纯度 85%），β-LgB（纯度>90%）和 β-LgA（纯度>90%）分别按不同的浓度梯度配制 6 种蛋白标准品溶液进行单个标准品标线的制定，每个浓度做 2 个平行，其中 R^2 分别达到 0.999 以上（表 3-1）。

表 3-1 蛋白标准品校准方程参数

Table 3-1 Parameters of calibration equations for single protein fractions

蛋白组分 Protein fraction	截距±标准误 Intercept±SE[a]	斜率±标准误 Slope±SE	R^2
α-CN	96 360.2±811.7	5 038.1±5.0	0.9993
β-CN	103 789.8±2 266.4	6 003.6±61.2	0.9995
κ-CN	5 748.8±358.8	137.0±0.7	0.9991
α-La	59 656.2±1 918.7	3 149.2±45.3	0.9990
β-LgA	33 567.5±836.2	3 474.2±22.9	0.9992

（续表）

蛋白组分 Protein fraction	截距±标准误 Intercept±SE[a]	斜率±标准误 Slope±SE	R^2
β-LgB	95 617. 8±729. 5	5 700. 3±26. 8	0. 9990

注：[a]标准误；

Note：[a] Standard error

3. 2. 6　数据分析

数据通过 Excel（2007）和 SAS 9. 1（SAS Institute Inc. ，Cary，NC）的 GLM 模块分别进行方差分析；并利用 Unscrambler 9. 8（CAMO SOFTWARE AS，Oslo，Norway）进行主成分分析。

3.3　结果

3. 3. 1　酪蛋白与乳清蛋白的分离

采用 RP-HPLC 法可在 30 min 分离乳中的酪蛋白及乳清蛋白。以标准品的出峰位置为基准确定乳样蛋白组分及其峰面积。在该检测条件下，经过单个蛋白标准品不同浓度的检测，确定 RP-HPLC 图出峰的顺序为：κ-CN，α-CN，α-La，β-CN_{A1}，β-CN_{A2}，β-LgB 和β-LgA，其中β-CN 标准品在保留时间为 17. 5 min 和 18 min 处均出现保留峰，且随着标准品浓度的升高，峰面积呈递增式扩大，故判断β-CN 标准品含异构体，将 17. 5 min 和 18 min 这两个出峰位点分别命名为β-CN_{A1}和β-CN_{A2}。在本试验样品中，奶牛和水牛乳蛋白检测到β-CN 的两种异构体，牦牛乳蛋白也检测出β-CN 的两种异构体，但其中的β-CN_{A2}含量较低（图 3-1）。

3. 3. 2　乳中总蛋白含量

奶牛、水牛、牦牛和山羊乳中的总蛋白含量采用凯氏定氮法测定，结果显示，四种畜种乳中的总蛋白含量差异显著，其中牦牛乳中的总蛋白含量最高，奶

牛乳中的总蛋白含量最低（表 3-2）。

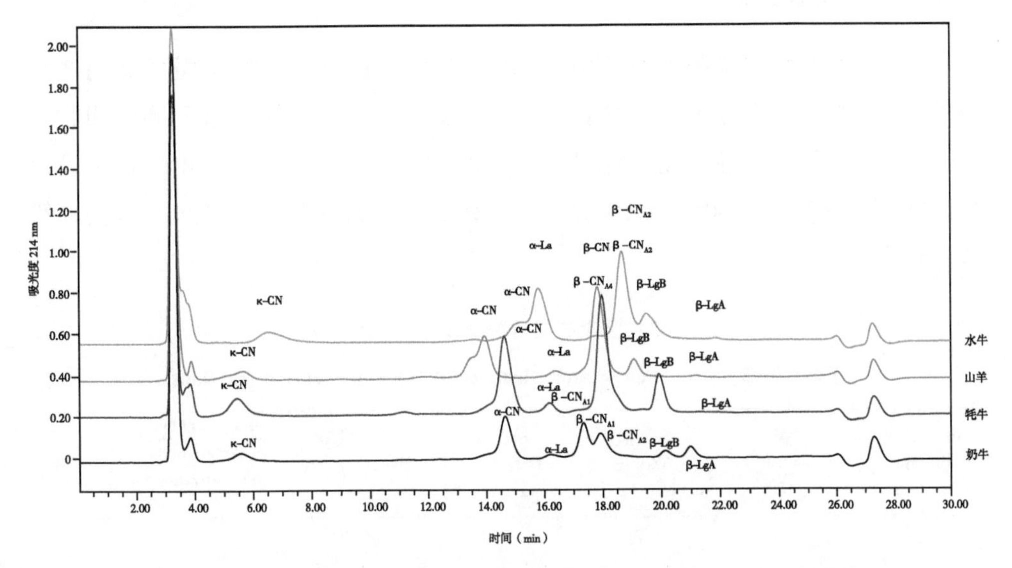

图 3-1　奶牛乳、水牛乳、牦牛乳和山羊乳蛋白 RP-HPLC 图谱

Fig. 3-1　Reversed phase high performance liquid chromatography chromatograms of protein fractions in buffalo, goat, yak and cow milk

表 3-2　凯氏定氮法测定奶牛乳、水牛乳、牦牛乳和山羊乳中总蛋白平均含量及标准偏差

Table 3-2　Mean and standard deviation (SD) of content (g/L) of milk total protein from cow, goat, buffalo and yak quantified by the Kjeldahl method

蛋白 Protein	物种 Species	平均含量 Mean（g/L）	标准偏差 SD
乳总蛋白	奶牛	37.35^{D}	0.818
	山羊	42.83^{C}	0.807
	水牛	56.09^{B}	0.405
	牦牛	62.87^{A}	1.009

注：$^{A-D}$肩标表示差异极显著（P<0.01）；

Note: * $^{A-D}$means within a row with different superscripts differ（P<0.01）

3.3.3 回收率测定

为了进一步确保检测的准确度，本试验进行了样品回收率的测定，进行 3 水平的加标试验，每个水平重复 3 次，根据峰面积计算检测值，进而测定回收率（表 3-3），乳蛋白主要组分的回收率为 90.53% ~ 97.78%，相对标准偏差（RSD）为 2.03%~9.90%，结果显示本测定方法的准确度和灵敏度均较高，符合实际检测要求。

表 3-3　乳蛋白组分检测结果准确度测定

Table 3-3　Accuracy of analytical results of milk protein fractions

蛋白组分 Protein fraction	含量 Concentration (g/L)	添加量 Adding amount (g/L)	检出量 Detectable amount (g/L)	回收率 Recovery rate (%)	平均回收率 Average recovery rate (%)	RSD (%)
α-CN	10.97	7.73	16.18	86.53	93.65	8.54
		15.49	24.37	92.11		
		23.23	34.99	102.31		
β-CN	12.98	8.67	20.08	82.71	90.53	8.72
		17.34	29.78	90.39		
		26.01	40.99	98.48		
κ-CN	5.91	3.41	10.66	86.51	92.33	6.47
		6.82	14.48	92.04		
		10.23	18.84	98.44		
α-La	1.39	0.61	2.11	89.96	97.78	9.90
		1.22	2.81	94.78		
		1.83	3.88	108.60		
β-LgA	2.02	1.2	3.39	88.30	91.39	4.83
		2.4	4.51	89.42		
		3.6	6.02	96.44		
β-LgB	1.08	1.27	2.56	89.18	91.15	2.03
		2.54	3.78	91.40		
		2.81	4.10	92.86		

3.3.4 酪蛋白和乳清蛋白的量化

奶牛乳、水牛乳、牦牛乳和山羊乳中主要蛋白组分含量见表 3-4，乳中主要蛋白组分含量均通过乳蛋白标准品的纯度进行折算获得。结果显示，较其他畜种乳，首先牦牛乳蛋白中的 κ-CN 含量最高（9.80 g/L）；其次是水牛乳、奶牛乳和山羊乳($P<0.01$)；最后牦牛乳中的 α-CN 含量亦极显著高于其他三种物种乳($P<0.01$)，但水牛乳和奶牛乳中 α-CN 的含量差异没有达到极显著（$P>0.01$），山羊乳中 α-CN 含量最低；牦牛和水牛乳中 β-CN 含量相近，且均显著高于山羊乳（$P<0.01$）。

山羊乳和水牛乳中的 β-LgB 含量相近，且均高于牦牛乳而低于奶牛乳($P<0.01$)；水牛乳中的 α-La 含量极显著地高于奶牛、牦牛和山羊乳（$P<0.01$），但奶牛、牦牛和山羊乳中的 α-La 含量差异不显著。奶牛乳中 β-LgA 含量极显著高于水牛、牦牛和山羊乳，但 β-LgA 在水牛、牦牛和山羊乳中的含量差异不显著。

表 3-4 奶牛乳、水牛乳、牦牛乳和山羊乳中主要蛋白组分平均含量及标准偏差

Table 3-4 Mean and standard deviation（SD）of contents（g/L）of the major milk protein fractions from cow，goat，buffalo and yak，quantified using calibration equations based on purified cow milk proteins

蛋白组分 Protein fraction	物种 Species	平均值（g/L） Mean（g/L）	标准偏差 SD
α-CN	奶牛	11.81^{B}	0.120
	山羊	9.48^{C}	0.258
	水牛	11.53^{B}	0.136
	牦牛	16.14^{A}	0.246
β-CN	奶牛	13.31^{C}	0.285
	山羊	20.53^{B}	0.490
	水牛	22.68^{A}	0.513
	牦牛	23.34^{A}	0.413

（续表）

蛋白组分 Protein fraction	物种 Species	平均值（g/L） Mean（g/L）	标准偏差 SD
κ-CN	奶牛	6.24^{C}	0.173
	山羊	5.56^{D}	0.053
	水牛	7.35^{B}	0.109
	牦牛	9.80^{A}	0.150
α-La	奶牛	1.45^{B}	0.032
	山羊	1.28^{B}	0.056
	水牛	7.96^{A}	0.156
	牦牛	1.47^{B}	0.038
β-LgA	奶牛	2.15^{A}	0.041
	山羊	0.14^{B}	0.006
	水牛	0.13^{B}	0.004
	牦牛	0.12^{B}	0.004
β-LgB	奶牛	1.19^{C}	0.031
	山羊	3.69^{B}	0.092
	水牛	3.88^{B}	0.052
	牦牛	9.46^{A}	0.227

注：$^{A-D}$肩标表示差异极显著（$P<0.01$）；

Note：$^{A-D}$ means within a row with different superscripts differ（$P<0.01$）

3.3.5 主成分（principal component analysis，PCA）分析

奶牛乳、水牛乳、牦牛乳和山羊乳中主要蛋白组分含量值通过主成分分析（PCA）软件 Unscrambler 9.8（CAMO Software AS，Oslo，Norway）进行多变量分析。PCA 分析的第一变量（PC1）和第二变量（PC2）的所得得分图（Scores）及载荷图（Losding）见图 3-2。在得分图（Scores）中距离相近的数据点表示相似，而距离较远的表示存在差异。由图 3-2 的结果得出，PC1 和 PC2 的得分图可以较好的区分奶牛、牦牛乳，水牛乳和奶牛乳及山羊乳和奶牛乳的区分效果亦较好，但水牛乳和山羊乳的区分效果不佳。由载荷图（Losding）的结果得出，

PC1和PC2的累计贡献率达到83.31%，其中PC1达到54.34%，PC2达到28.97%。将载荷图叠加到得分图可以确定每个物种所对应的主要蛋白组分，即奶牛乳中β-LgA高于其他物种，水牛乳中α-La含量极显著高于其他物种，而牦牛乳中则是α-CN、κ-CN和β-LgB含量均高于其他物种，与统计分析结果一致。

3.4 讨论

在本试验条件下，采用RP-HPLC测定奶牛乳、水牛乳、牦牛乳和山羊乳蛋白，为了解以上物种主要乳蛋白组分含量的差异以及建立本试验条件下的RP-HPLC图谱提供一定的基础。Hallén等（2007）的研究表明，奶牛乳蛋白中β-CN存在多种基因型，分别为A1、A2、A3和B型[159]。而Bonfatti等（2013）的研究指出，在水牛乳中，乳蛋白β-CN只有一种基因型[32]。早在1963年的研究就表明，水牛乳中乳蛋白α-CN、β-CN和κ-CN之间的比例有别于奶牛乳[56]。Feligini等（2009）通过RP-HPLC检测水牛乳蛋白的结果发现，在a_{S2}-CN和a_{S1}-CN之间存在一个峰，经过质谱鉴定为κ-CN的一种异构体，但没有进一步进行分析其他的特征[29]。Mercier等（1975）对山羊乳蛋白κ-CN的主要结构进行了检测，结果发现，在山羊乳中κ-CN只有一种结构，没有发现κ-CN的异构体[160]。本试验的结果显示，除了没有在山羊乳中检测到κ-CN的异构体外，其他的物种乳蛋白中也没有检测出κ-CN的异构体。在本试验条件下，α-La的出峰位置位于α-CN和β-CN之间，β-LgB的出峰位置位于β-CN之后而在β-LgA之前。此外，在山羊乳中只检测到β-CN的一个峰，出峰保留时间为18 min，与Trujillo等（2000）的报道一致[57]。不同物种乳蛋白中的乳清蛋白β-LgA和β-LgB呈现不同的色谱图，与之前的相关研究报道一致，Ferreira和Caçote（2003）通过购置Sigma Chemical Co.的牛科类乳蛋白β-Lg和α-La标准品进行奶牛、山羊和绵羊乳蛋白的分离检测，结果显示，不同物种的主要乳清蛋白组分在RP-HPLC呈现不同的色谱图[161]。此外，前人通过疏水作用色谱法和RP-HPLC，采用Sigma（St. Louis，MI，USA）公司提供的蛋白标准品进行不同畜种乳蛋白的分离检测，结果均显示不同畜种乳蛋白呈现不同的色谱峰[162-163]。此外，其他的一

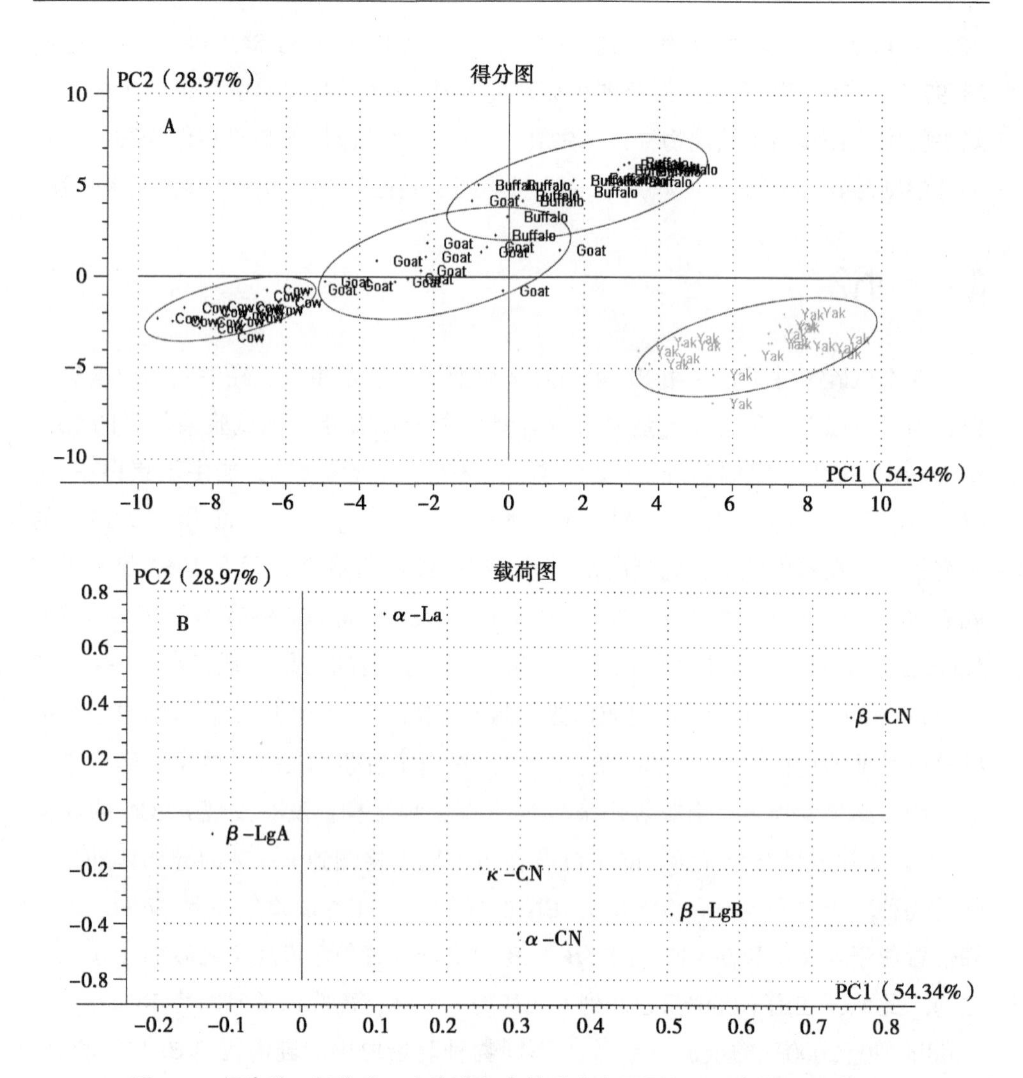

图 3-2　奶牛乳、牦牛乳、水牛乳和山羊乳中主要蛋白组分含量 PCA 分析

Fig. 3-2　Principal component analysis（scores plot A and loadings plot B）showing differences in composition of protein fractions in milk from four species measured by RP-HPLC：red color，cow；green color，goat；blue color，buffalo；light blue，yak

些因素，例如，低 pH 值引起的钙流失可能会导致 α-La 折叠构象和疏水性的改变[164]，这种现象是否会引起其他蛋白构象的改变有待于进一步的研究。在本试

验中，没有进一步分析遗传及构象差异所引起的一些可能的色谱图差异，因此，后期的研究有待于进一步进行深入。本试验中，成功地应用 RP-HPLC 分离和量化了奶牛、水牛、牦牛和山羊乳主要乳蛋白组分，为快速检测不同物种乳蛋白提供了一种方法。不同物种乳主要乳蛋白组分的 RP-HPLC 图谱不同，这将为区分不同物种乳或是不同来源乳提供一种检测方法。

有研究通过毛细管电泳技术检测超高温灭菌乳（ultra-high-temperature，UHT），脱脂乳和豆乳中的乳清蛋白 α-La 和 β-Lg，并认为该方法用来检测和分析乳中的 α-La 和 β-Lg 比较简单[49,165-166]，而本试验是通过 RP-HPLC 检测乳中主要蛋白组分，包括乳清蛋白 α-La 和 β-Lg，且结果显示，牦牛乳中主要的蛋白组分含量均最高，除 β-LgA 之外，而水牛乳中，乳清蛋白 α-La 的含量最高，显著高于奶牛乳，与前人的报道相一致[30]。Hallén 等（2008）采用 HPLC 分析检测奶牛乳中的主要蛋白组分及其基因型，结果显示，β-Lg 的含量为 5.20 g/L，较本试验的结果高（3.34 g/L），而 α-La 的含量则比本试验的检测结果低（0.98 g/L vs. 1.45 g/L）[159]。但本试验的结果与 Marchi 等（2010）采用中红外光谱预测的奶牛乳中乳清蛋白 β-Lg 和 α-La 含量的结果相一致[167]。Bonfatti 等（2008）的研究结果显示，奶牛乳蛋白中乳清蛋白 α-La 和 β-Lg 的比例为 1∶3[31]，较本试验的结果高。Gasilova 等（2012）通过毛细管电泳技术检测 UHT 乳中的 α-La 和 β-Lg 的含量，结果显示，α-La 的含量为 0.15 mg/mL±0.01 mg/mL，β-Lg 的含量为 0.52 mg/mL±0.03 mg/mL[166]，均稍低于本试验的检测结果，可能是由于 α-La 和 β-Lg 经过超高温处理后部分变性所致，但具体原因有待于进一步研究探讨。本试验的检测结果显示，水牛乳中 α-La 的含量显著高于 Bonfatti 等（2013）[32] 和 Buffonia 等（2011）[30] 的研究报道，而 β-Lg 的含量与 D’Ambrosio 等（2008）的研究结果相近（7.1 g/kg vs. 7.7 g/kg）[79]，但其中的 β-LgB 的含量与 Buffonia 等（2011）的结果相比较低（3.88 g/L vs. 4.04 g/L）[30]。水牛乳蛋白中的乳清蛋白 β-Lg 的含量不是 α-La 含量的 1.3 倍，这与 Bonfatti 等（2013）的研究结果不一致[32]，可能与水牛本身的遗传基因型有关。测定所得的牦牛乳蛋白中的乳清蛋白 α-La 和 β-Lg 含量均高于 Li 等（2010）的研究结果[38]，且接近于其测定的最高值，但 β-LgA 的含量则低于 Li 等（2010）

的研究结果[38]。

奶牛乳蛋白的组成及含量的鉴定早有报道[168-169]。Marchi 等（2010）通过中红外预测奶牛乳蛋白组分含量结果显示，乳蛋白中的酪蛋白含量与本试验的结果存在一定的差异，其中总酪蛋白、α-CN 和 β-CN 的含量均比本试验的结果高（35.10 g/L vs. 31.36 g/L、16.86 g/L vs. 11.81 g/L、14.79 g/L vs. 13.31 g/L），但 κ-CN 的含量低于本试验的研究结果（3.71 g/L vs. 6.24 g/L）[167]。但 Hallén 等（2008）的研究结果显示，其通过 RP-HPLC 测定的奶牛乳蛋白中的总酪蛋白、α-CN、β-CN 和 κ-CN 的含量均略低于本试验的结果（24.89 g/L vs. 31.36 g/L、9.97 g/L vs. 11.81 g/L、11.85 g/L vs. 13.31 g/L、3.07 g/L vs. 6.24 g/L）[159]。Bonfatti 等（2011）测定散装牛乳的结果显示，乳中的总蛋白、酪蛋白和乳清蛋白的含量分别为 40.57 g/L、35.38 g/L 和 5.19 g/L，均略高于本试验的研究结果[170]。

Feligini 等（2009）通过 RP-HPLC 测定水牛乳蛋白含量的结果显示，水牛乳中的 α_{s1}-CN、α_{s2}-CN、β-CN 和 κ-CN 含量分别为 8.89 g/L、5.08 g/L、20.91 g/L 和 4.13 g/L，其中 β-CN 的含量与本试验的研究结果相近（20.91 g/L vs. 22.68 g/L），α-CN 的含量则略高，而 κ-CN 的含量则略低于本试验的研究结果[29]。与 Bonfatti 等（2013）的研究结果相比，本试验测定的总乳蛋白含量略高（56.09 g/L vs. 60.10 g/L），而 κ-CN 的含量相近（7.35 g/L vs. 7.79 g/L），α-CN 的含量则显著高于 Bonfatti 等（2013）的测定结果（11.53 g/L vs. 24.14 g/L），β-CN 的含量则显著低于 Bonfatti 等（2013）的测定结果（22.68 g/L vs. 18.45 g/L）[32]，可能与水牛的品种等因素有关，具体原因有待于进一步研究探讨。

有关山羊乳蛋白组分含量等的研究报道指出，山羊乳中总蛋白含量为 36.7 g/L±2.6 g/L，平均酪蛋白含量为 29.7 g/L±2.3 g/L，其中 α_{s1}-CN、β-CN、α_{s2}-CN 和 κ-CN 分别为 21.8 g/kg、43.8 g/kg、13.7 g/kg 和 13.8 g/kg[36]，本试验的结果显示，山羊乳中总蛋白含量和酪蛋白含量为 42.83 g/L 和 35.37 g/L，较 Moatsou 等（2006）的研究结果略高，但 α-CN 含量略低（26.7 g/kg vs. 35.5 g/kg），β-CN 和 κ-CN 的含量略高（57.7 g/kg vs. 43.8 g/kg、15.6 g/kg vs. 13.8

g/kg)[36]。山羊乳蛋白中酪蛋白β-CN 是主要的蛋白组分，与 Park（2008）的研究结果相一致[171]。

有关牦牛乳蛋白含量的研究报道较少，Li 等（2010）的研究指出，牦牛乳蛋白中α-CN 的含量为 16.03 g/L，与本研究的结果相一致（16.14 g/L），但β-CN 和κ-CN 的含量略低（18.2 g/L vs. 23.34 g/L、5.98 g/L vs. 9.80 g/L）[38]，且本试验的结果显示，牦牛乳蛋白中的β-CN 和κ-CN 含量均与 Li 等（2010）研究结果的最大值相一致[38]。

3.5 小结

（1）RP-HPLC 被认为是一种快速分离和鉴定乳蛋白组分的一种方法，在本试验中成功地分离和量化了奶牛、山羊、水牛及牦牛乳蛋白组分。

（2）不同畜种（奶牛、山羊、水牛及牦牛）乳蛋白组分含量存在较大差异，奶牛乳中β-LgA 显著高于其他畜种，水牛乳中α-La 含量显著高于其他畜种，牦牛乳中的α-CN、β-CN 及β-LgB 含量显著高于其他畜种，且可通过 PCA 可较为明显的区分不同畜种乳。

（3）在本试验条件下，建立了奶牛乳、山羊乳、水牛乳和牦牛乳的 RP-HPLC 特征图谱，由于在 RP-HPLC 呈现不同的色谱图，这为进一步区分不同来源乳，甚至掺假乳提供一定的理论基础。

4 基于GC-MS技术分析不同乳畜乳中OBCFA及其组分含量

4.1 引言

在大多数的植物中，OBCFA 组成的含量一般都是微量级别的[99]。然而，在一些动物乳和组织中，OBCFA 的含量和组成是有所变化的，如奶牛[100]、绵羊[101]、山羊[102]和海狸[103]。Keeney 等（1962）研究指出，乳中的 OBCFA 大部分是来自瘤胃微生物[104]，且近期的研究亦证明该结论[94,105]。奶牛乳中 OBCFA 主要包含 *iso*-C13:0（tridecanoic acid isomers）、*iso*-C14:0（tetradecanoic acid）、C15:0，*iso*-C15:0 和 *anteiso*-C15:0（pentadecanoic acid）、*iso*-C16:0（hexadecanoic acid）、C17:0、*iso*-C17:0 和 *anteiso*-C17:0（heptadecanoic acid）[107]。而山羊乳中的 BCFA 的检测亦有报道，且对奶牛和山羊乳中 BCFA 的差异性进行了检测[108]。此外，DePooter 等（1981）研究指出，尽管奶牛与山羊乳中具有奇数或偶数碳原子的顺式脂肪酸的数量相似，但其含量仍存在较大的差异[109]。反刍动物和单胃动物乳中的脂肪及其组成存在较大的差异，在反刍动物乳中，绵羊乳中的 OBCFA 含量最多，占总 FA 的 5.5%，且在反刍动物与单胃动物乳脂肪中，占据主导地位的 OBCFA 组分为 C15:0 和 C17:0[110]。有关牦牛乳[111-113]、娟姗牛乳[101,114-115]、马乳[116]和骆驼乳[117-118]中脂肪的研究均有报道。然而，有关牦牛乳、骆驼乳、马乳、水牛乳和娟姗牛乳中OBCFA的相关数据报道较有限，且介于这些物种乳中 OBCFA 的组成含量的比较研究没有。此外，随着对非牛科类哺乳动物乳需求量的不断增长[76]。有关不同物种乳中脂肪组成和功能，包括 OBCFA 等的相关研究有待于进一步深入。

本试验通过从奶牛乳、水牛乳、牦牛乳、娟姗牛乳、山羊乳、骆驼乳和马乳

中萃取出 OBCFA，通过 GC-MS 进行测定，以检测不同物种乳中 OBCFA 的组分含量及其差异性，从而进一步了解不同物种乳脂肪特征。此外，奶牛乳、水牛乳、牦牛乳、娟姗牛乳、山羊乳、骆驼乳和马乳中的 OBCFA 组分含量数据均通过主成分 PCA 和聚类分析，进一步检测和区分不同物种乳。本试验测定的乳中的 OBCFA 包括 *anteiso*-C13:0、*iso*-C14:0、C15:0，*iso*-C15:0 和 *anteiso*-C15:0、*iso*-C16:0、C17:0，*iso*-C17:0 和 *anteiso*-C17:0。

4.2 试验材料与方法

4.2.1 样品采集

于 2013 年 2 月至 4 月，从北京某养殖场采集奶牛（105 d±15 d）乳样 20 份，从河北某养殖场采集山羊（120 d±21 d）乳样 20 份，从河北某养殖场采集娟姗牛（107 d±20 d）乳样 20 份，从云南某养殖场采集水牛（115 d±20 d）乳样 20 份，从青海某养殖场采集牦牛（110 d±12 d）乳样 20 份，从新疆某养殖场采集马（110 d±16 d）乳样和骆驼（100 d±22 d）乳样各 8 份，乳样于-20℃条件下保存，并利用装有干冰的泡沫箱运送至中国农业科学院北京畜牧兽医研究所，于-80℃条件下保存至分析。所有供体试验动物均处于健康状态，均处于泌乳日龄约为 4 个月。

4.2.2 脂肪酸标准品

乳中脂肪酸的测定采用的标准品如下：isoC14:0、C15:0，iso-C15:0 和*anteiso*-C15:0、C17:0 和 iso-C17:0 的标准品均采购自 Sigma-Aldrich（USA）；*anteiso*-C13:0、iso-C16:0 和 *anteiso*-C17:0 的标准品均采购自 Laordan AB（Sweden）。

4.2.3 试剂

4.2.3.1 $NaOCH_3$/Methanol（氢氧化钠甲醇溶液）

将 2 g NaOH 溶于 100 mL 含水不超过 0.5%的甲醇中，溶液放置一段时间后，可能产生 Na_2CO_3 白色沉淀而失效，此时重新配制。溶液冷藏保存。

4.2.3.2 盐酸/甲醇溶液

10 mL 乙酰氯慢慢加入到 100 mL 无水甲醇中，冷藏保存。

4.2.3.3 Na_2SO_4溶液

6.67 g Na_2SO_4 溶于 100 mL 纯水中。

4.2.3.4 正己烷/异丙醇混合溶液

正己烷/异丙醇以 3∶2 比例混合，冷藏保存。

4.2.3.5 正己烷

色谱纯级正己烷。

4.2.4 样品的前处理

样品的前处理参考 Bu 等（2007）[172]的方法。

（1）将乳样于4℃解冻 5 h 后，取 2 mL 乳样加 4 mL 正己烷/异丙醇溶液，再加 Na_2SO_4 溶液 2 mL，室温离心 5 300 r/min，20 min。

（2）提取上清液在 20 mL 水解管中，混合后氮气吹干。

（3）加入 2 mL $NaOCH_3$/Methanol 在 50℃水浴 15 min，冷却后加入 2 mL 盐酸/甲醇溶液在 80℃水浴 1.5 h。

（4）冷却到室温，加入 3 mL 水和 6 mL 正己烷，震荡，静置或离心分层。

（5）吸取上层液体（尽量吸净），氮气吹干，加入 1 mL 正己烷，旋涡振荡 1 min，将溶液吸取至 2 mL 离心管中，10 000 r/min 离心 2 min，取上清液上机测定。

4.2.5 仪器设备

气相色谱质谱联用仪（gas chromatography-mass spectrometry in scan detection mode，GC-MS/SCAN）（Agilent，7890A-5975B，America），带有 EI 离子源。

4.2.6 气相色谱条件

（1）色谱柱：HP－88（J&W Scientific，Folsom，CA，USA），100 m×

0. 25 mm×0. 25 μm。

（2）柱温：120℃保持10 min，1. 5℃ /min升温至230℃，保持35 min。

（3）载气：氮气（纯度99. 999%），流速1. 0 mL/min，恒流模式。

（4）进样量：1 μL，不分流进样。

（5）进样口温度：250℃。

采用外标法定量。

4. 2. 7 质谱条件

MS接口温度：280℃；EI温度：70 ev；离子源温度：230℃；四极杆温度：150℃；分析模式：Scan（全离子扫描）。

4. 2. 8 数据分析

数据采用SAS 9. 1（SAS Institute Inc. , Cary, NC）进行统计分析及邓肯多重比较分析。OBCFA所有组分的数据进一步通过Unscrambler 9. 8（CAMO SOFTWARE AS, Oslo, Norway）和Gene Cluster3. 0（USA）中的Average Linkage进行主成分分析（PCA）和聚类分析（Cluster）。

4.3 结果

4. 3. 1 不同畜种乳中OBCFA的含量测定分析

奶牛乳、水牛乳、牦牛乳、娟姗牛乳、山羊乳、骆驼乳和马乳中OBCFA组分含量及总量如表4-1和表4-2及图4-1和图4-2（g/100 g FA）所示。不同畜种乳中的OBCFA均采用CG-MS进行测定。在四种牛科物种：奶牛乳、水牛乳、牦牛乳和娟姗牛乳中OBCFA组分中含量最高的是*iso*-C15∶0和C15∶0，而山羊乳中OBCFA组分中含量最高的是C15∶0和*anteiso*-C17∶0，在马乳和骆驼乳中OBCFA组分中含量最高的是*iso*-C15∶0和*anteiso*-C17∶0。在上述所有不同畜种乳中，*anteiso*-C13∶0的含量最低。由表4-1和表4-2的结果可以得出，牦牛乳中的 OBCFA 总含量最高，且牦牛乳中 OBCFA 组分中除 *anteiso*-C17:0 外，其他组分的含量亦高于其他物种，而 *anteiso*-C17:0 的含量则是略低于骆驼乳。而马乳中的总 OBCFA 含量及其组分的含量均显著低于其他物种。

奶牛乳、水牛乳、牦牛乳、娟姗牛乳、山羊乳、骆驼乳和马乳中 OBCFA 组分含量如表4-3所示（g/100 g OBCFA），其结果与统计单位为 g/100 g FA 的结果相似。

表 4-1 不同物种乳中 OBCFA 组分含量（g/100 g FA）

Table 4-1 Odd and branched chain fatty acid (OBCFA) profile in milk from different species (g/100 g FA)

物种 Species	奶牛 Cow	牦牛 Yak	水牛 Buffalo	娟姗牛 Jersey cattle	山羊 Goat	骆驼 Camel	马 Horse
anteiso-C13:0	0.007± 0.0002^{Ef}	0.052± 0.0012^{Aa}	0.047± 0.0018^{ABb}	0.016± 0.0007^{De}	0.026± 0.0011^{Cd}	0.042± 0.0024^{Bc}	0.009± 0.0009^{Ef}
iso-C14:0	0.074± 0.0030^{Ee}	0.682± 0.0104^{Aa}	0.335± 0.0147^{Bb}	0.251± 0.0079^{Cc}	0.171± 0.0066^{Dd}	0.321± 0.0099^{Bb}	0.063± 0.0049^{Ee}
anteiso-C15:0	0.049± 0.0016^{Cc}	0.295± 0.0075^{Aa}	0.171± 0.0053^{Bb}	0.173± 0.00631^{Bb}	0.080± 0.0042^{Cc}	0.185± 0.0138^{ABb}	0.004± 0.0003^{Cc}
iso-C15:0	1.042± 0.0275^{DEe}	3.348± 0.0431^{Aa}	2.901± 0.1298^{Bb}	3.0707± 0.1112^{ABab}	1.370± 0.0363^{Dd}	2.268± 0.1287^{Cc}	0.813± 0.0871^{Ee}
C15:0	0.699± 0.0198^{Ee}	3.320± 0.0439^{Aa}	2.190± 0.0477^{Bb}	1.981± 0.0303^{Cc}	1.942± 0.0563^{CDc}	1.768± 0.0659^{Dd}	0.308± 0.0153^{Ff}
iso-C16:0	0.183± 0.0057^{Dd}	1.270± 0.0312^{Aa}	0.493± 0.0192^{BCb}	0.5343± 0.0234^{Bb}	0.400± 0.0406^{Cc}	0.516± 0.0282^{Bb}	0.137± 0.0112^{Dd}
anteiso-C17:0	0.636± 0.0222^{Ff}	2.220± 0.0304^{Bb}	1.086± 0.0215^{Ee}	1.415± 0.0665^{Dd}	2.000± 0.0683^{Cc}	2.892± 0.1334^{Aa}	0.784± 0.0430^{Ff}
iso-C17:0	0.348± 0.0119^{Ff}	1.362± 0.0156^{Aa}	0.616± 0.0136^{Dd}	0.789± 0.0305^{Cc}	0.535± 0.0198^{Ee}	1.021± 0.0220^{Bb}	0.213± 0.0125^{Gg}
C17:0	0.616± 0.0244^{Dd}	3.020± 0.0618^{Aa}	1.139± 0.0352^{Cc}	1.198± 0.0391^{Cc}	1.503± 0.0329^{Bb}	1.412± 0.0369^{Bb}	0.280± 0.0214^{Ee}

注：$^{a-g}$肩标表示差异显著（P<0.05），$^{A-G}$肩标表示差异极显著（P<0.01）；

Note: $^{a-g}$means within a row with different superscripts differ (P<0.05), $^{A-G}$ means within a row with different superscripts differ (P<0.01)

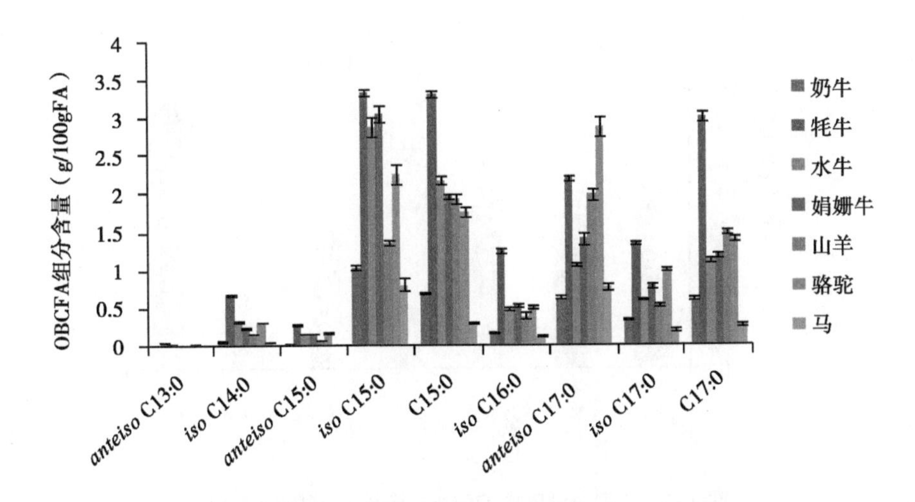

图 4-1　不同畜种乳中 OBCFA 组分含量（g/100 g FA）

Fig. 4-1　Odd and branched chain fatty acid（OBCFA）profile in milk from different species（g/100 g FA）

表 4-2　不同畜种乳中 OBCFA 含量（g/100 g FA）

Table 4-2　Total odd and branched chain fatty acid（OBCFA）composition in milk from different species（g/100 g FA）

物种 Species	奇数碳链支链脂肪酸 OBCFA	标准偏差 SD
娟姗牛	9.43^{C}	0.20
水牛	8.98^{C}	0.21
骆驼	10.42^{B}	0.34
奶牛	3.66^{E}	0.08
山羊	8.03^{D}	0.12
马	2.61^{F}	0.13
牦牛	15.57^{A}	0.14

注：$^{A-F}$肩标表示差异极显著（P <0.01）；

Note：$^{A-F}$means within a row with different superscripts differ（P<0.01）

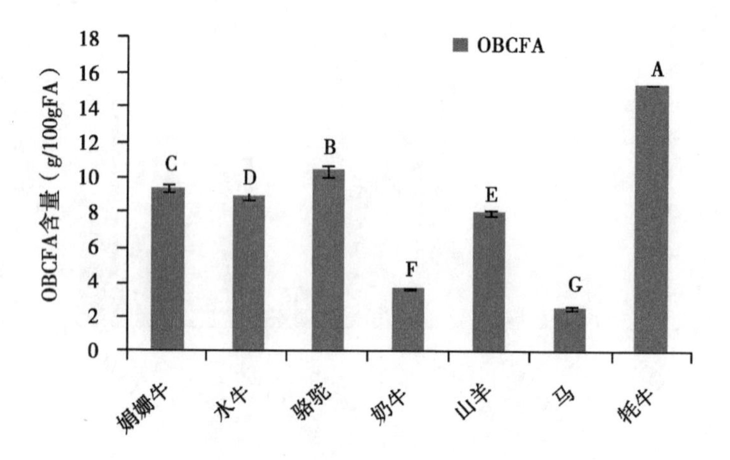

图 4-2 不同畜种乳中 OBCFA 含量（g/100 g FA）

Fig. 4-2 Total odd and branched chain fatty acid（OBCFA）composition in milk from different species（g/100 g FA）

表 4-3 不同畜种乳中 OBCFA 组分含量（g/100 g OBCFA）

Table 4-3 Odd and branched chain fatty acid（OBCFA）compositions of milk from different species milk（g/100 g OBCFA）

物种 Species	奶牛 Cow	牦牛 Yak	水牛 Buffalo	娟姗牛 Jersey cattle	山羊 Goat	骆驼 Camel	马 Horse
anteiso-C13:0	0.0020± 0.0001^{Def}	0.0034± 0.0001^{Cc}	0.0052± 0.0002^{Aa}	0.0019± 0.00005^{Df}	0.0023± 0.0002^{De}	0.0039± 0.0002^{Bb}	0.0029± 0.0002^{Cd}
iso-C14:0	2.02± 0.0658^{Dd}	4.40± 0.0413^{Aa}	3.71± 0.1080^{Bb}	2.69± 0.1254^{Cc}	1.75± 0.0876^{Dd}	2.80± 0.1298^{Cc}	2.51± 0.1305^{Cc}
anteiso-C15:0	1.35± 0.0416^{Bb}	1.90± 0.0438^{Aa}	1.88± 0.0336^{Aa}	0.98± 0.0280^{Cc}	0.93± 0.0326^{Cc}	1.91± 0.0946^{Aa}	0.11± 0.0157^{Dd}
iso-C15:0	28.79± 0.5671^{Bb}	21.63± 0.2740^{Cc}	32.42± 0.7915^{Aa}	32.98± 1.0194^{Aa}	19.13± 1.0484^{Cc}	21.82± 0.6935^{Cc}	29.73± 1.7626^{ABb}
C15:0	19.36± 0.47478^{Cd}	21.42± 0.1384^{Cc}	24.15± 0.3880^{Bb}	20.59± 0.5269^{Ccd}	26.55± 1.1653^{Aa}	15.59± 0.6158^{De}	12.13± 0.4684^{Ef}
iso-C16:0	5.04± 0.1421^{CDc}	8.18± 0.1357^{Aa}	5.40± 0.2115^{Cc}	6.39± 0.1681^{Bb}	4.46± 0.2111^{Dd}	5.35± 0.1093^{Cc}	5.64± 0.2670^{BCc}
anteiso-C17:0	17.04± 0.3370^{Dd}	14.33± 0.1326^{Ee}	11.99± 0.2176^{Ff}	14.50± 0.5597^{Ee}	21.91± 0.7269^{Cc}	28.95± 1.0262^{Bb}	33.94± 0.5880^{Aa}

（续表）

物种 Species	奶牛 Cow	牦牛 Yak	水牛 Buffalo	娟姗牛 Jersey cattle	山羊 Goat	骆驼 Camel	马 Horse
iso-C17:0	9.78± 0.2566^{Aa}	8.80± 0.0819^{ABb}	6.80± 0.1298^{Dd}	8.22± 0.2944^{Bbc}	7.10± 0.2966^{CDd}	9.62± 0.2680^{Aa}	7.88± 0.2682^{BCc}
C17:0	16.28± 0.4095^{Bb}	19.00± 0.2071^{Aa}	12.59± 0.2545^{Cc}	12.88± 0.5035^{Cc}	18.53± 0.5331^{Aa}	13.37± 0.3102^{Cc}	10.48± 0.5240^{Dd}

注：$^{a-f}$肩标表示差异显著（$P<0.05$），$^{A-F}$肩标表示差异极显著（$P<0.01$）；

Note: Values are means±SD, $^{a-f}$ means within a row with different superscripts differ ($P<0.05$), $^{A-F}$ means within a row with different superscripts differ ($P<0.01$)

4.3.2 不同畜种乳中 OBCFA 的 PCA 分析

通过 PCA 对不同畜种乳中 OBCFA 各组分含量（g/100 g FA）进行分析得出（图 4-3A），马乳和奶牛乳在 PC1 上聚为一组，牦牛乳在 PC1 上可以很好地与其他畜种乳进行区分，而骆驼乳和山羊乳则可较好的在 PC1 和 PC2 中间与其他畜种进行区分，娟姗牛乳和水牛乳在 PC1 方向聚为一组，结果显示不同畜种乳中 OBCFA 存在一定的差异性。总体来讲，PCA 可以很好地将牦牛乳和其他畜种乳进行区分，而马乳、奶牛乳、骆驼乳和山羊乳的区分效果较好，但通过 PCA 无法将水牛乳和娟姗牛乳进行区分。不同畜种乳中 OBCFA 的 PCA 分析的载荷图见图 4-3B，该图的每一个点分别对应不同畜种乳中相应的 OBCFA 组分，因此可以通过该图上的点所对应的 OBCFA 组分对不同畜种乳进行区分。PCA 前两个主成分变量累计贡献率达 90.48%，其中 PC1 为 79.09%，PC2 为 11.39%，该贡献率值已满足其作为一个参考指标进行不同畜种乳的区分。PC1 方向上主要含有 *iso*-C15:0、C17:0 和 C15:0，而 PC2 方向上主要含有 *iso*-C15:0 和 *anteiso*-C17:0。

通过 PCA 对不同畜种乳中 OBCFA 各组分含量（g/100 g OBCFA）进行分析的结果见图 4-4，其中前两个主成分的得分图见图 4-4A。鉴于距离相近的点表示相似性较强，距离稍远的点表示相似性较弱，即存在一定的差异性，由图 4-4A的研究结果得出，水牛乳、娟姗牛乳、奶牛乳、牦牛乳和山羊乳具有一定的相似性而聚为一类。马乳和骆驼乳中的 OBCFA 较相似而聚为一类。PCA 分析的前两个主成分累计达到 83.09%，其中 PC1 为 44.85%，PC2 为 38.24%。采用

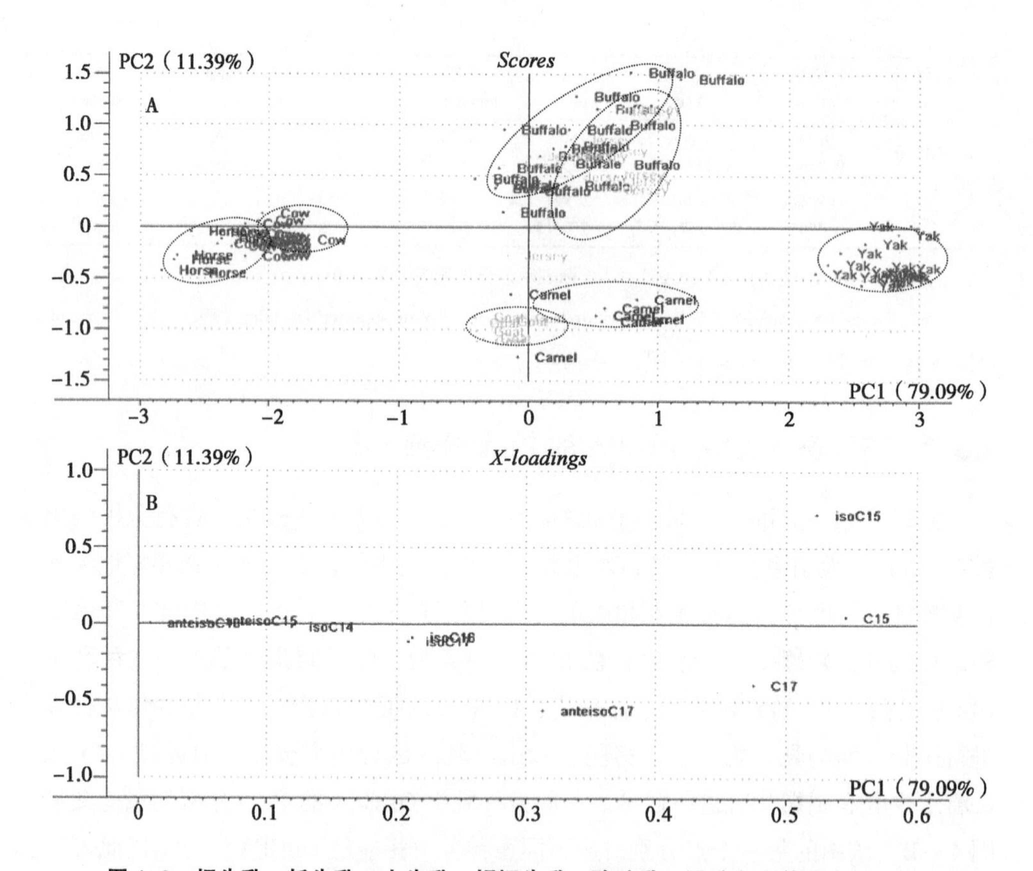

图 4-3　奶牛乳、牦牛乳、水牛乳、娟姗牛乳、骆驼乳、马乳和山羊乳中 OBCFA 组分含量 PCA 分析（g/100 g FA）

Fig. 4-3　Principal component analysis (scoring plot A, and loading plot, B) showing the odd and branched chain fatty acid (OBCFA) profiles in milk from different species as measured by gas chromatography-mass spectrometry (GC-MS): Red: buffalo, Green: cow, Blue: camel, Pink: yak, Brown: horse, Watchet blue: goat and Gray: Jersey cattle (g/100 g FA)

PCA 分析以 g/100 g OBCFA 为统计单位的乳中 OBCFA 组分含量的结果与以 g/100 g FA 为统计单位的乳中 OBCFA 组分含量的结果略有差异，可能是由于不同畜种乳中总的 OBCFA 含量存在差异的缘故。

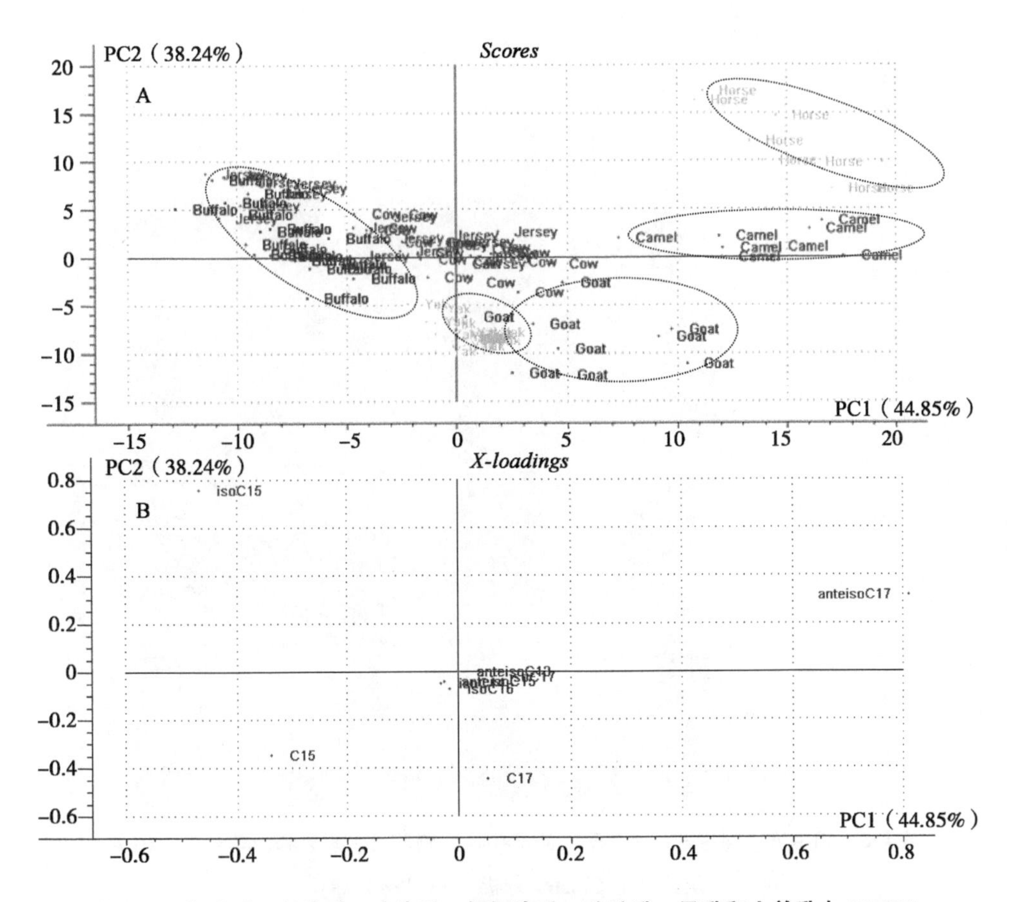

图 4-4 奶牛乳、牦牛乳、水牛乳、娟姗牛乳、骆驼乳、马乳和山羊乳中 OBCFA 组分含量 PCA 分析（g/100 g OBCFA）

Fig. 4-4 Principal component analysis (scoring plot A, and loading plot, B) showing odd and branched chain fatty acid (OBCFA) profiles of milk from different species, measured by GC-MS, Red: camel, Green: cow, Blue: buffalo, Pink: yak, Brown: horse, Watchet blue: goat and Gray: Jersey cattle (g/100 g OBCFA)

4.3.3 不同畜种乳中 OBCFA 的 Cluster 分析

不同畜种乳中 OBCFA 含量（g/100 g FA）采用 Cluster 分析的结果见图 4-5。

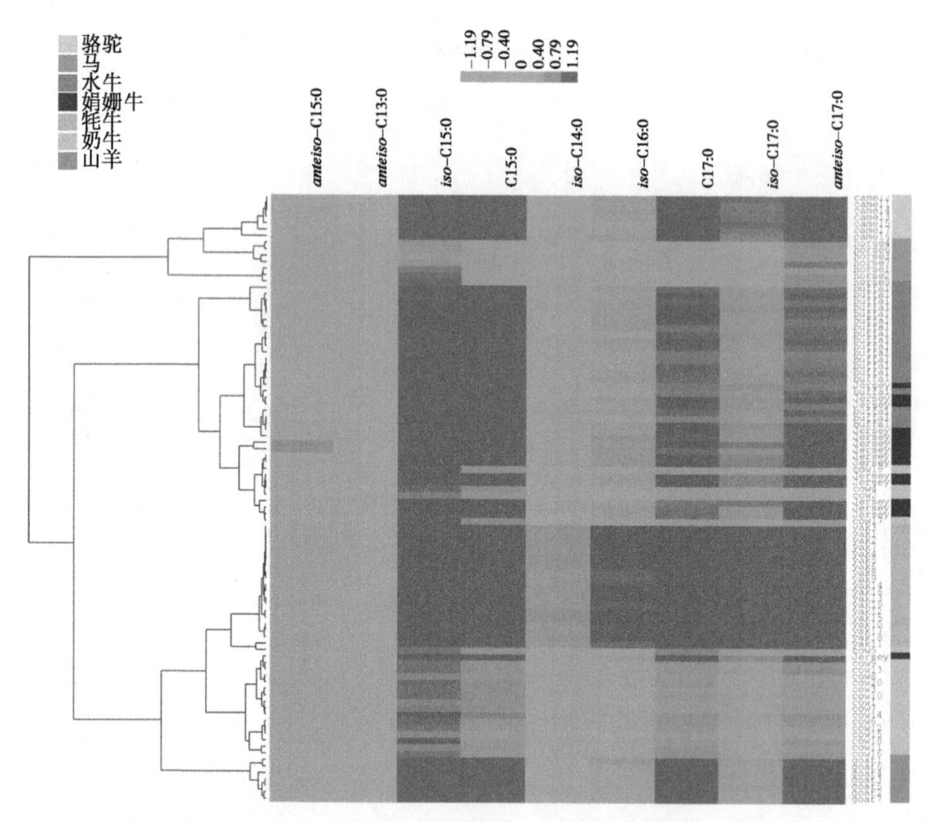

图 4-5　奶牛乳、牦牛乳、水牛乳、娟姗牛乳、骆驼乳、马乳和山羊乳中 OBCFA 组分含量 Cluster 分析（-1. 19~1. 19）

Fig. 4-5　Hierarchical clustering of odd and branched chain fatty acids（OBCFAs） for different species' milk. The clustering was based on the individual OBCFA compositions of camel，cow，buffalo，yak，horse，goat and Jersey cattle milk. The bar color represents a logarithmic scale from -1. 19 to 1. 19（g/100 g FA）

以上物种主要聚为两大类，其中四种牛科类：奶牛乳、水牛乳、牦牛乳和娟姗牛乳聚为一类，山羊乳也聚在这一类，但又独立聚一小类；而骆驼乳和马乳聚为另一大类，具有一定的种属特异性，与统计分析的结果相一致。该结果与 Yang 等（2013）通过蛋白质组学技术分析不同畜种乳清中蛋白组分而进行聚类分析的结果相一致[53]。且通过聚类分析，除娟姗牛乳外，其他畜种乳均可较好的进行区分。采用 GC-MS 技术对奶牛乳、水牛乳、牦牛乳、娟姗牛乳、山羊乳、骆驼乳和马乳中 OBCFA 进行测定得出，不同畜种乳中 OBCFA 组分含量差异显著，牦牛乳中 OBCFA 总量和其组分除 *anteiso*-C17:0 外，含量均最高。在四种牛科类：奶牛乳、水牛乳、牦牛乳和娟姗牛乳 OBCFA 中含量最高的是 *iso*-C15:0 和 C15:0；而在山羊乳则是 C15:0 和 *anteiso*-C17:0；马和骆驼乳中则是 *iso*-C15:0 和 *anteiso*-C17:0，均呈现一个物种特异性。通过 PCA 和 Cluster 均可较好的区分奶牛乳、水牛乳、牦牛乳、娟姗牛乳、山羊乳、骆驼乳和马乳，且 Cluster 的结果与统计分析结果相一致，聚类效果具有一定的物种特异性。鉴于聚类分析的结果可以得出，除了娟姗牛外，其他畜种乳均可通过 OBCFA 组分进行很好的区分，同样的结论适用于以 g/100 g OBCFA 为统计单位的 Cluster 分析结果（图 4-6）。

4.4 讨论

结果显示，四种牛科类（奶牛、牦牛、水牛和娟姗牛）乳中的 OBCFA 组分中含量最高的是 *iso*-C15:0 和 C15:0；山羊乳中含量最高的 OBCFA 组分是 C15:0 和 *anteiso*-C17:0；而马乳和骆驼乳中的含量最高的 OBCFA 组分是 *iso*-C15:0 和 *anteiso*-C17:0。这些结果与 Devle 等（2012）的研究报道几乎一致，他指出，在反刍动物与非反刍动物乳中 OBCFA 含量最高的组分均为 C15:0 和 C17:0[110]。Yang 等（2000）的研究报道指出，OBCFA 主要组分对人类抗癌作用比较明显，其可以较为显著地抑制癌细胞的生长[173]。而大量的研究指出，在 OBCFA 中主要起抗癌作用的 OBCFA 组分为 *iso*-C15:0 和 *anteiso*-C15:0[93-94,97,173]。该结果对人类抗癌治疗将起一定的促进作用。此外，马乳中的 OBCFA 组分含量较低，可能与马后肠中缺少合成这些组分的微生物所致[110]。

在反刍动物乳中发现了一些特殊的 BCFA，包括山羊、绵羊和奶牛，且奶牛乳通常含有高比例的含有多于或少于 16 个碳原子的脂肪酸[174]。此外，有些研究指出，奶牛乳中主要的 BCFA 为 *iso*-和 *anteiso*-类脂肪酸[94,105,175]。当以 g/100 g OBCFA 为统计单位时，Lima（2011）比较研究了古巴奶牛和 Vlaeminck 等

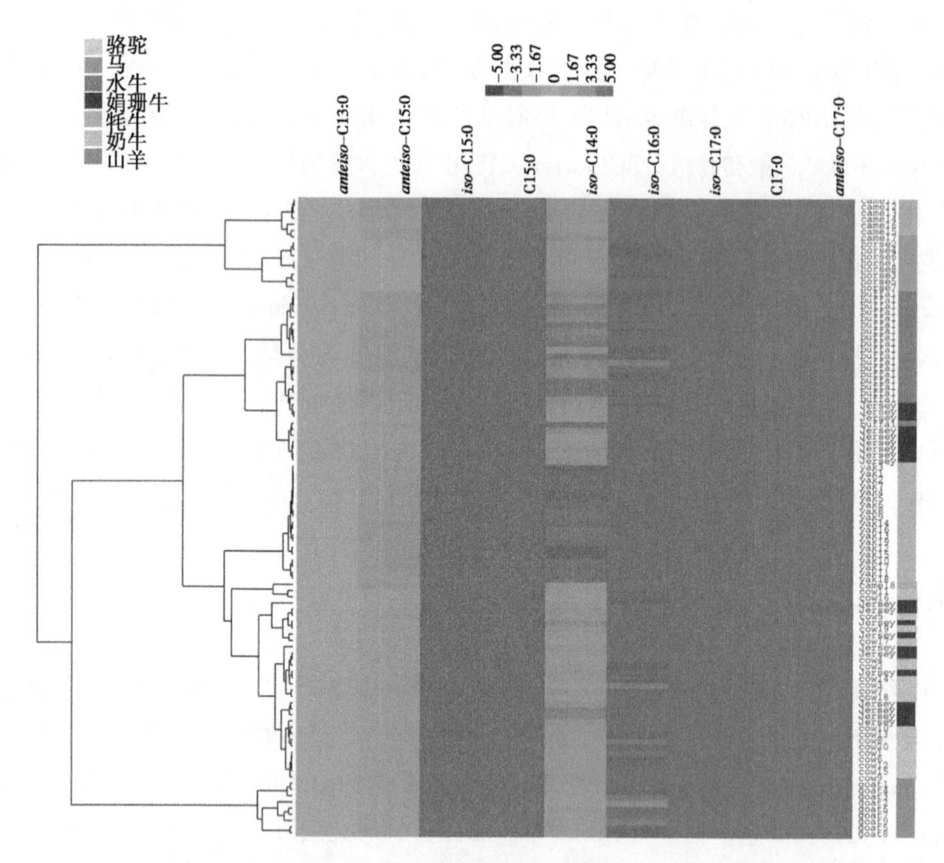

图 4-6 奶牛乳、牦牛乳、水牛乳、娟姗牛乳、骆驼乳、马乳和山羊乳中 OBCFA 组分含量 Cluster 分析（-5.0~5.0）

Fig. 4-6 Hierarchical clustering of odd and branched chain fatty acids (OBCFAs) of milk from different species. The clustering was based on individual OBCFA compositions in camel, cow, buffalo, yak, horse, goat, Jersey cattle. The bar color represents a logarithmic scale from -5.0 to 5.0 (g/100 g OBCFA)

（2006）报道的荷斯坦奶牛乳中 OBCFA，结果显示，不同品种的奶牛乳中的 OBCFA 的组成含量存在一定的差异；且 Lima（2011）的研究结果指出，奶牛乳中含量最高的 OBCFA 组分为 C15:0，与本试验的研究结果相一致，但 *anteiso*-C15:0的含量在 Lima（2011）的研究结果中占据第二位[176]，该结果与本试验结

果存在一定的差异。有关山羊乳中 OBCFA 的研究报道较少。马乳中 OBCFA 含量的研究亦较少，Devle 等（2012）报道的马乳中 OBCFA 含量较本试验的研究结果较低，但奶牛乳中的 OBCFA 含量与本试验的结果相近[110]。此外，本试验结果通过 Cluster 分析结果显示，其中四种牛科类：奶牛乳、水牛乳、牦牛乳和娟姗牛乳聚为一类，山羊乳也聚在这一类，但又独立聚一小类；而骆驼乳和马乳聚为另一大类，具有一定的种属特异性，与 Yang 等（2013）通过蛋白质组学技术研究分析的奶牛乳、水牛乳、牦牛乳、山羊乳和骆驼乳，及聚类分析的结果相一致[53]。

此外，鉴于前人报道 OBCFA 组分有抗癌作用，在本试验条件下，通过 GC-MS 技术对奶牛乳、水牛乳、牦牛乳、娟姗牛乳、山羊乳、骆驼乳和马乳中 OBCFA 含量的测定，为乳中 OBCFA 的实际应用提供一定的数据基础，为乳脂肪的进一步研究提供一定基础信息。

4.5 小结

（1）在奶牛乳、水牛乳、牦牛乳和娟姗牛乳脂肪 OBCFA 组成中含量最高的均是*iso*-C15:0和 C15:0；而在山羊乳中是 C15:0 和 *anteiso*-C17:0；在马和骆驼乳中是 *iso*-C15:0 和 *anteiso*-C17:0，均呈现一定的种属特异性。

（2）PCA 可以很好地将牦牛乳和其他物种乳进行区分，对马乳、奶牛乳、骆驼乳和山羊乳的区分效果较好，但通过 PCA 无法将水牛乳和娟姗牛乳进行区分。

（3）通过 Cluster 聚类分析不同物种乳中 OBCFA 的结果得出，以上物种主要聚为两大类，其中四种牛科类：奶牛乳、水牛乳、牦牛乳和娟姗牛乳聚为一类，山羊乳也聚在这一类，但又独立聚一小类；而骆驼乳和马乳聚为另一大类，具有一定的种属特异性。

5 基于ICP-MS方法分析不同畜种乳中常量及微量元素含量

5.1 引言

乳中含有大量的可以满足人类需要的营养物质，如蛋白、脂类、碳水化合物、维生素和元素等。作为一种组成复杂的生物活性物质，乳中富含大量的常量元素和微量元素，参与机体大量的生物生化反应过程，在满足人类机体营养需要方面起着重要的作用。有关乳中矿物质元素的研究报道较多，鉴于乳中元素除了在人体营养方面起一定的生物学功能，而其中的某些元素对机体有一定的毒副作用，因此，研究乳中元素的含量及组成是非常有必要的[119]。然而，乳中常量及微量元素的含量变异性较大，主要影响因素较多，其中与物种本身的基因遗传性、乳腺的分泌、健康状况及泌乳阶段均有关系[120]。

随着检测技术的不断发展，检测的灵敏度及准确度也在不断地提高。截至目前，已有较多的关于奶牛乳中元素测定的研究报道，且亦有一些研究采用的是ICP-MS 技术[119-120,123-127]。Khan 等（2006）通过分光光度计测定结果显示，山羊和绵羊乳中的元素 Ca、K、Mg、Na 和 Zn 含量存在显著的差异性[128]。Güler（2007）通过采用 ICP-MS 测定山羊乳中 24 种元素的含量[129]。但有关骆驼和马乳中元素含量的报道较少，且没有同时测定比较分析奶牛乳、水牛乳、牦牛乳、水牛乳、骆驼乳和马乳中元素含量的研究报道。

ICP-MS 检测技术作为具有高敏感度和高选择性，以及可同时分析多种元素的一种快速的标准检测技术已运用于很多研究领域[33,130-133]。在本研究中，采用 ICP-MS 技术检测奶牛乳、水牛乳、牦牛乳、娟姗牛乳、山羊乳、骆驼乳和马乳中 9 种元素，其中包括 4 种常量元素和 5 种微量元素，而这些乳样均需经过微波

消解这一程序，该程序已被认为是检测食品中元素含量的一种非常有效的方法。同时，采用购自国家标准物质中心的两种生物成分分析标准物质，奶粉（GBW10017）和圆白菜（GBW10014）作为标准物质进行ICP-MS测定过程中质量控制的检测。

此外，由于采用多变量数据分析可获得较多的信息，且鉴于已有研究，根据食品中的常量和微量元素组成和含量，将模型识别技术应用于估计食品质量安全。此外，通过最优技术分析奶牛乳中的常量及微量元素可为其提供一种进行样品归类的方法，进而为进一步通过元素区别不同来源乳。在本研究中，采用PCA分析奶牛乳、水牛乳、牦牛乳、娟姗牛乳、山羊乳、骆驼乳和马乳中的9种元素（K、Ca、Na、Mg、Zn、Fe、Mn、Co和Se），进而根据物种来源进行乳样的区分，为区分不同来源乳提供一定的依据。

5.2 试验材料与方法

5.2.1 样品采集

于2013年2月至4月，从北京某养殖场采集奶牛（105 d±15 d）乳样20份，从河北某养殖场采集山羊（120 d±21 d）乳样20份，从河北某养殖场采集娟姗牛（107 d±20 d）乳样20份，从云南某养殖场采集水牛（115 d±20 d）乳样20份，从青海某养殖场采集牦牛（110 d±12 d）乳样20份，从新疆某养殖场采集马（110 d±16 d）乳样和骆驼（100 d±22 d）乳样各8份，乳样于-20℃条件下保存，并利用装有干冰的泡沫箱运送至中国农业科学院北京畜牧兽医研究所，于-80℃条件下保存。所有供体试验动物均处于健康状态，均处于泌乳日龄约为4个月。

5.2.2 试剂

5.2.2.1 硝酸和过氧化氢

BVⅢ级的HNO_3（北京化工）和H_2O_2（30%）（Merck，Germany）用于样品

消解，以破坏样品中有机物。此外，硝酸还用于浸泡器皿和配制标准溶液与标准曲线。

5.2.2.2 元素标准品

锰（Mn）、铁（Fe）、钴（Co）、铜（Cu）、锌（Zn）和硒（Se）元素标准准备液购买自 Agilent 公司，浓度均为 10 mg/L（Agilent Part#: 8 500~6 940）。用 1%HNO_3 将其稀释 100 倍后作为工作液，浓度为 100 ng/mL；钾（K）（GBW（E）080125）、钙（Ca）、（GBW（E）080118）钠（Na）[GBW（E）080127] 和镁（Mg）[GBW（E）080126] 元素标准储备液购自国家标准物质中心，浓度均为1 000 ug/mL。

5.2.3 标准曲线的制备

采用外标法对所测定的样品进行定量测定分析。标准曲线各元素的标准品分别吸取不同体积，4%HNO_3 定容于 50 mL。定容后标准曲线浓度分别为钾（K）、钙（Ca）、钠（Na）和镁（Mg）：0、1 ng/mL、2 ng/mL、3 ng/mL、4 ng/mL、5 ng/mL，6 个梯度浓度；锰（Mn）、钴（Co）和硒（Se）：0、0.5 ng/mL、1 ng/mL、1.5 ng/mL、2 ng/mL，5 个梯度浓度；铁（Fe）、钴（Co）、铜（Cu）和锌（Zn）：0、50 ng/mL、100 ng/mL、150 ng/mL、200 ng/mL，5 个梯度浓度。

5.2.4 仪器设备

微波消解仪为 MARS 5（CEM，美国），配有 50 mL 聚四氟乙烯消解管。电感耦合等离子体质谱仪（ICP-MS）为 Agilent 7700X（Agilent，美国），高频率（3 MHz）双曲面四级杆。配有 I-AS 89 位的自动进样器，Perltier 同心雾化器，HMI 高基体的进样系统，双通路的 Scott 雾化室。

5.2.5 样品消解与测定条件

5.2.5.1 样品消解

将不同畜种乳样品解冻后充分混匀，吸取 1 mL 乳样于微波消解罐内，加入 2 mL 的 HNO_3，5 mL 的 H_2O_2，盖上内盖，旋紧外盖。置于微波消解仪中，进行

梯度升温消解。微波消解程序见表 5-1。同时进行空白试验。消解后冷却至 60℃ 左右，用超纯水定容至 50 mL，待测。

表 5-1　微波消化操作程序

Table 5-1　Operating program for the microwave digestion

步骤 Step	功率 Power （W）	梯度温度时间 Gradient temperature time（min）	温度 Temperature （℃）	保持时间 Holding time（min）
1	1 600	10	90	6
2	1 600	10	140	10
3	1 600	10	190	20

5.2.5.2　仪器测定条件

ICP-MS 的测定条件见表 5-2。

表 5-2　ICP-MS 设备条件及参数

Table 5-2　ICP-MS Instrument condition and data acquisition parameters

参数 Parameter	条件 Experimental conditions
雾化器 Nebulizer	同心喷雾器 Concentric Nebulizer
雾化室 Spray chamber	石英双通道斯科特喷雾室 Quartztwo-channel Scott Spray Chamber
雾化室温度 Spray chamber temperature	2℃ 2℃
射频 RF power	1 550 W 1 550 W
进样深度 Smpl depth	8 mm 8 mm
样品提升量 Sample uptake rate	0.15 mL/min 0.15 mL/min
载气 Carrier gas flow rate	0.9 L/min 0.9 L/min
等离子体气 Plasma gas flow rate	15.0 L/min 15.0 L/min

（续表）

参数 Parameter	条件 Experimental conditions
辅助气 Auxiliary gas flow rate	0. 25 L/min 0. 25 L/min
反应气 He gas	5 mL/min 5 mL/min

5. 2. 6 数据分析

数据采用 SAS 9. 1（SAS Institute Inc. , Cary, NC）进行统计分析及邓肯多重比较分析。OBCFA 所有组分的数据进一步通过 Unscrambler 9. 8（CAMO SOFTWARE AS, Oslo, Norway）进行主成分分析（PCA）。

5.3 结果

5. 3. 1 测定过程的质量控制

采用购自国家标准物质中心生物成分分析标准物质——奶粉（GBW10017）和生物成分分析标准物质——圆白菜（GBW10014）作为标准物质进行 ICP-MS 测定过程中质量控制的检测，该标准物质的消解过程及测定条件均与样品处理过程相同。具体结果见表 5-3 和表 5-4。所有测定的结果显示，标准物质的各元素含量与测定值之间无显著差异，均在其真值不确定度范围内，表明所采用的方法及处理过程在测定奶牛乳、水牛乳、牦牛乳、娟姗牛乳、山羊乳、骆驼乳和马乳中元素含量的结果准确可靠。

表 5-3 ICP-MS 测定准确性检测（奶粉，GBW10017）

Table 5-3 Validation of the accuracy of the ICP-MS multielemental procedure proposed（milk powder, GBW10017）

元素 Element	检定值 Certified value（μg/g）	测定值 Observed value[a]（μg/g）
Na	47±3	45. 9±0. 2

（续表）

元素 Element	检定值 Certified value（μg/g）	测定值 Observed value[a]（μg/g）
Mg	9.6±0.7	9.1±0.1
K	125±5	121.5±1.8
Ca	94±3	91.9±2.9
Mn	5.1±1.7	4.9±0.1
Fe	78±13	84.5±1.9
Co	0.3±0.07	0.3±0.01
Cu	5.1±1.3	4.9±0.4
Zn	340±20	353±3.1
Se	1.1±0.3	1.2±0.1

注：[a] n=3

表 5-4　ICP-MS 测定准确性检测（圆白菜，GBW10017）

Table 5-4　Validation of the accuracy of the ICP-MS multielemental procedure proposed（Cabbage，GBW10014）

元素 Element	检定值 Certified value（μg/g）	测定值 Observed value[a]（μg/g）
Na	109±6	113.4±3
Mg	24.1±1.5	25±1.3
K	155±6	154.8±5.4
Ca	70±2	69.7±1.8
Mn	187±8	190.2±1.5
Fe	980±100	967.5±6.5
Co	8.9±1.4	9.0±0.6
Cu	27±2	25.8±1
Zn	260±20	257.6±2
Se	2±0.3	2.2±0.1

注：[a] n=3

5.3.2　不同畜种乳中常量元素的测定

奶牛乳、水牛乳、牦牛乳、娟姗牛乳、山羊乳、骆驼乳和马乳中的 4 种常量元素（Na、Mg、K和Ca）含量见表5-5，由表5-5的结果可以看出，骆驼乳中的Na的含量最高，而Mg元素含量较低；马乳中的4种常量元素Na、Mg、K和Ca的含量较其他物种均最低，且差异显著；山羊乳中的4种常量元素含量均较高，且K的含量是所有畜种乳中含量最高的；水牛乳中的Mg和Ca元素含量最高，而奶牛、牦牛和娟姗牛乳中除Ca和Na元素外，其他的2种常量元素（Mg、K）含量差异不显著。

表5-5 奶牛乳、水牛乳、牦牛乳、娟姗牛乳、山羊乳、骆驼乳和马乳中常量元素含量（mg/L）

Table 5-5 Comparative major elemental concentrations in cow, yak, buffalo, Jersey cattle, goat, camel and horse milk（mg/L）

元素 Element	奶牛 Cow	牦牛 Yak	水牛 Buffalo	娟姗牛 Jersey cattle	山羊 Goat	骆驼 Camel	马乳 Horse
Na	318. 07± 17. 63^{Dd}	319. 04± 10. 50^{Dd}	380. 94± 12. 57^{CDc}	390. 10± 16. 58^{Cc}	488. 83± 27. 04^{Bb}	579. 61± 22. 73^{Aa}	155. 10± 2. 93^{Ee}
Mg	120. 44± 3. 32^{Ccd}	110. 91± 3. 51^{Cd}	182. 00± 8. 72^{Aa}	130. 90± 3. 58^{Cc}	154. 77± 5. 94^{Bb}	81. 13± 2. 71^{De}	64. 42± 2. 74^{Df}
K	1 320. 29± 26. 30^{B}	1 361. 60± 22. 40^{B}	918. 26± 16. 86^{C}	1 412. 52± 35. 75^{B}	1 546. 06± 58. 93^{A}	844. 81± 41. 03^{C}	307. 91± 17. 50^{D}
Ca	1 350. 38± 45. 18^{B}	1 046. 97± 56. 95^{C}	1 821. 53± 33. 36^{A}	1 373. 87± 32. 50^{B}	1 379. 51± 51. 60^{B}	1 426. 18± 63. 01^{B}	670. 04± 38. 98^{D}

注：$^{a-f}$肩标表示差异显著（P<0. 05），$^{A-E}$肩标表示差异极显著（P<0. 01）；

Note：$^{a-f}$ means within a row with different superscripts differ（P<0. 05），$^{A-E}$ means within a row with different superscripts differ（P<0. 01）

5. 3. 3 不同畜种乳中微量元素的测定

奶牛乳、水牛乳、牦牛乳、娟姗牛乳、山羊乳、骆驼乳和马乳中的5种微量元素（Mn、Co、Zn、Fe和Se）含量见表5-6，由表5-6的结果可以看出，马乳中的Mn元素较其他畜种含量最低，且差异显著；奶牛乳、牦牛乳、娟姗牛乳、骆驼乳和马乳中的Co元素含量较山羊和水牛乳中的Co元素含量高，且差异显著，但山羊乳和水牛乳中Co元素含量差异不显著；奶牛乳、水牛乳和骆驼乳中的Zn元素含量较其他畜种乳高；奶牛乳中的Fe元素含量最高，稍高于牦牛乳，而显著高于其他畜种乳；Se元素在水牛乳中含量最高，在骆驼乳中含量最低，

且极显著的低于其他畜种乳，但在马乳中没有检测到 Se 元素。

表 5-6 奶牛乳、水牛乳、牦牛乳、娟姗牛乳、山羊乳、骆驼乳和马乳中微量元素含量（μg/L）

Table 5-6 Comparative trace elemental concentrations in cow，yak，buffalo，Jersey cattle，goat，camel and horse milk（μg/L）

元素 Element	奶牛 Cow	牦牛 Yak	水牛 Buffalo	娟姗牛 Jersey cattle	山羊 Goat	骆驼 Camel	马乳 Horse
Mn	73. 80± 3. 18^{Aa}	46. 27± 2. 36^{Cc}	39. 31± 2. 25^{Ccd}	38. 65± 1. 81^{Ccd}	60. 28± 3. 12^{Bb}	36. 68± 2. 52^{Cd}	16. 81± 1. 87^{De}
Co	0. 97± 0. 05^{A}	1. 02± 0. 07^{A}	0. 37± 0. 03^{B}	0. 96± 0. 07^{A}	0. 32± 0. 02^{B}	0. 92± 0. 07^{A}	1. 02± 0. 10^{A}
Zn	7 352. 35± 379. 02^{Aa}	5 814. 94± 295. 69^{Bc}	6 323. 35± 214. 35^{ABbc}	5 409. 81± 178. 69^{Bc}	4 144. 73± 264. 62^{Cd}	7 066. 66± 452. 44^{Aab}	4 113. 36± 146. 17^{Cd}
Fe	1 591. 22± 91. 21^{Aa}	1 182. 72± 140. 17^{ABb}	776. 78± 30. 10^{BCcd}	535. 75± 43. 30^{Ccd}	439. 88± 20. 78^{Cd}	862. 65± 73. 90^{BCbc}	633. 75± 43. 93^{Ccd}
Se	18. 45± 1. 22^{B}	12. 63± 0. 52^{C}	27. 34± 1. 19^{A}	18. 11± 0. 66^{B}	14. 34± 0. 88^{C}	1. 22± 0. 23^{D}	n. d.

注：$^{a-d}$肩标表示差异显著（P<0. 05），$^{A-D}$肩标表示差异极显著（P<0. 01）；

Note：$^{a-d}$ means within a row with different superscripts differ（P<0. 05），$^{A-D}$ means within a row with different superscripts differ（P<0. 01）

5. 3. 4 不同畜种乳中元素（Na、Mg、K、Ca、Mn、Co、Zn、Fe 和 Se）含量的 PCA 分析

将奶牛乳、水牛乳、牦牛乳、娟姗牛乳、山羊乳、骆驼乳和马乳中的 9 种元素（Na、Mg、K、Ca、Mn、Co、Zn、Fe 和 Se）含量通过主成分 PCA 进行分析，结果见图 5-1。PCA 分析的前三个主成分累计达到 98. 40%，其中 Z：PC1 为 88. 73%，X：PC2 为 28. 46%，Y：PC3 为 1. 21%。由 PCA 得分图（图 5-1A）可知，娟姗牛乳在 PC1 上聚为一类，可以很好地与其他畜种乳进行区分，而马乳在 PC3 上聚为一组，可以很好地与其他畜种乳进行区分。结合 PCA 载荷图（图 5-1B）可知，马乳中因缺少硒元素而与其他乳区分明显，水牛乳中的钙、山羊奶中的钾以及娟姗牛乳中铁使得其可以和其他乳较为明显的区分。

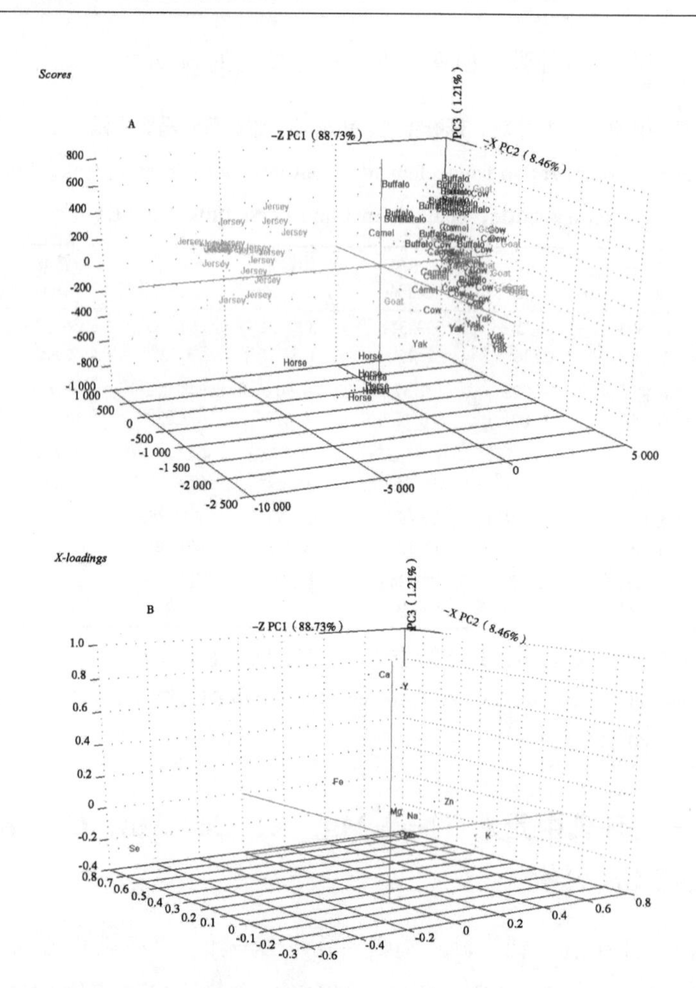

图 5-1　奶牛乳、牦牛乳、水牛乳、娟姗牛乳、骆驼乳、马乳和山羊乳中常量及微量元素含量 PCA 分析（红色：骆驼；绿色：奶牛；蓝色：水牛；棕色：马；浅绿色：山羊；灰色：娟姗牛；粉色：牦牛）

Fig. 5-1　Principal component analysis（scores plot A，and loadings plot，B）showing trace and minor elements in milk of different species measured by inductively coupled plasma-mass spectrometry（ICP-MS）：Red colour，camel；Green colour，cow；Blue colour，buffalo；Brown，horse；Watchet blue，goat；Gray，Jersey cattle；Pink，yak

5.4 讨论

结果显示，奶牛乳、水牛乳、牦牛乳、娟姗牛乳、山羊乳、骆驼乳和马乳中的 Na、Mg、K、Ca、Mn、Co、Zn、Fe 和 Se 含量差异显著。Khan 等（2006）研究分析了山羊乳和绵羊乳中的 Na、Mg、K、Ca、Mn、Co、Zn、Fe、Cu 和 Se 的含量，结果显示绵羊乳中的 Mg、K、Na 和 Fe 的含量均高于山羊乳，但 Co 和 Se 的含量差异不显著，而山羊乳中的 Ca、Cu、Mn 和 Zn 含量显著高于绵羊[128]。此外，Rodríguez 等（2001）测定山羊乳中的八种元素，其平均含量分别为 Se：19.98 μg/L、Fe：0.52 mg/L、Cu：0.17 mg/L、Zn：3.31 mg/L、Na：514 mg/L、K：1 585mg/L、Ca：1 533 mg/L、Mg：157 mg/L，其中的 Se、Fe、Zn、Na、K、Ca 和 Mg 的含量与本试验所测定的山羊乳中以上元素含量的结果相近[177]。Sola-Larrañaga 等（2009）通过 ICP-MS 技术分析西班牙纳瓦拉地区奶牛乳中的元素，结果显示，奶牛乳中的 Ca 的平均含量为 970 mg/L，Mg 的平均含量为 91.8 mg/L，Na 的平均含量为 372 mg/L，K 的平均含量为 1 344 mg/L，Zn 的平均含量为 4 631 μg/L，Fe 的平均含量为 290 μg/L，Mn 的平均含量为 29.1 μg/L，Se 的平均含量为 9.77 μg/L，其中 Na 的含量较本试验的研究结果偏高（372 mg/L vs. 318.07 mg/L），K 的含量与本试验的研究结果相近（1 344 mg/L vs. 1 320.29 mg/L），其他的元素含量均低于本试验的研究结果[120]。Rahimi（2013）研究比较了伊朗不同地区山羊乳、奶牛乳、绵羊乳和水牛乳中的 Ca 的含量，结果显示，奶牛乳中 Ca 的平均含量为 0.92 ng/mL±0.47 ng/mL，水牛乳中 Ca 的平均含量为 0.74 ng/mL±0.41 ng/mL，山羊乳中 Ca 的平均含量为 2.33 ng/mL±1.23 ng/mL，绵羊乳中的 Ca 的平均含量为 3.31 ng/mL±1.53 ng/mL，其中奶牛和水牛乳中 Ca 的含量均低于本试验的研究结果，而山羊乳中的 Ca 的含量高于本试验的研究结果[178]，这可能与物种本身的遗传特异性有关，具体原因有待于进一步研究。Barlowska（2011）和 Medhammar 等（2012）的研究报道指出[6,11]，奶牛乳中的 Ca 的含量为 113 mg/100 g，Na 的含量为 43 mg/100 g，K 的含量为 132 mg/100 g，Mg 的含量为 10 mg/100 g，Fe 的含量为 30 μg/100 g，Zn 的含量为 400 μg/100 g，其中常量元素 Ca、Na、K 和 Mg 的含量均与本试验的研究结果相近，而微量元素 Fe 和 Zn 的含量均低于本试验的研究结果；水牛乳中 Ca 的含量为 191 mg/100 g，Na 的含量为 47 mg/100 g，K 的含量为 112 mg/100 g，Mg 的含量为 12 mg/100 g，Fe 的含量为 170 μg/100 g，Zn 的含量为 500 μg/100 g，其中 Ca 的含量与本试验的研究结果相近，而其他的元素含量均略低于本试验的研究结果；牦牛乳中 Ca 的含量为 129 mg/100 g，Na 的含量为 29 mg/100 g，K 的含量为 95 mg/100 g，Mg 的含量为 10 mg/100 g，Fe 的含量为 570 μg/100 g，Zn 的含量为 900 μg/100 g，其中常量元素除 K 的含量低于本试验的研究结果外，其余的均与本试验结果相近，而微量元素 Fe 和 Zn 的含量均高于本试验的研究结果；山羊乳中 Ca 的含量为 132~134 mg/100 g，Na 的含量为 41~59.4 mg/100 g，K 的含量为 152~181 mg/100 g，Mg 的含量为 15.8~16 mg/100 g，Fe 的含量为 7~60 μg/100 g，Zn 的含量为 56~370 μg/100 g，其中的常量元素 K、Ca、Na 和 Mg 及微量元素 Fe 的含量均与本试验的研究结果相一致，但微量元素 Zn 的含量低于本试验的研究结果；骆驼乳中 Ca 的含量为 114~116 mg/100 g，Na 的含量为 59 mg/100 g，K 的含量为 144~156 mg/100 g，Mg 的含量为 10.5~12.3 mg/100 g，Fe 的含量为 230~290 μg/100 g，Zn 的含量为 530~590 μg/100 g，其中常量元素 K 和 Mg 的含量均高于本试验的研究结果，而 Na 与本试验的结果相近，但 Ca 的含量略低于本试验的结果，微量元素 Fe 的含量显著高于本试验的检测结果，而 Zn 的含量则较低；马乳中 Ca 的含量为 95 mg/100 g，Na 的含量为 16 mg/100 g，K 的含量为 51 mg/100 g，Mg 的含量为 7 mg/100 g，Fe 的含量为 100 μg/100 g，Zn 的含量为 200 μg/100 g，其中元素 Ca、K 和 Fe 的含量均略高于本试验的结果，Na 和 Mg 的含量与本试验的结果相一致，元素 Zn 的含量显著低于本试验的结果。然而，由于本试验的研究是基于比较不同物种乳中几种元素的差异性，对各物种的日粮结构以及饲养环境等因素没有进行评估，故在后续的研究中有必要进一步对以上因素对各畜种乳中元素含量的影响进行深入研究。

5.5 小结

（1）奶牛乳、水牛乳、牦牛乳、娟姗牛乳、山羊乳、骆驼乳和马乳中的 4 种

常量元素 Na、Mg、K、Ca 和 5 种微量元素 Mn、Co、Zn、Se、Fe 含量差异显著。

（2）马乳中的 9 种元素，其中包括 4 种常量元素（Na、Mg、K、Ca）和 5 种微量元素（Mn、Co、Zn、Se、Fe）含量均显著低于其他畜种乳；水牛乳中的 Mg 和 Ca 的含量较高；而奶牛乳中的 Fe 的含量最高；骆驼和山羊乳中的 Na 和 K 含量显著高于其他物种。

（3）除 Se 的含量略低于水牛乳以外，奶牛乳中的其他微量元素（Mn、Co、Zn、Se）含量均高于水牛乳、牦牛乳、娟姗牛乳、山羊乳、骆驼乳和马乳；除在山羊乳中 Co 元素的含量略低外，奶牛乳、水牛乳、牦牛乳、娟姗牛乳、山羊乳、骆驼乳和马乳中的 Co 含量相近；而马乳中未检测出 Se。

6 基于Label-Free定量蛋白质组学方法分析不同畜种乳MFGM蛋白

6.1 引言

乳脂是牛乳的重要成分，是由乳腺上皮细胞内质网合成的，以甘油三酯为中心外包有复杂的脂质双分子层膜。当脂滴释放进入牛乳时，乳脂球中的不同细胞质成分保留到不同的膜层中[60]。因此，乳脂主要由中性脂质、胆固醇、极性脂质和蛋白混合物组成[61]。尽管 MFGM 蛋白仅占牛乳总蛋白的 1%~4%，但与其他牛乳成分相比具有更复杂的多样性[62]。由于乳脂球的功能与营养特性，许多研究对它们的蛋白成分进行了分析[63-64]。首先，MFGM 在牛乳水相脂质中起到了乳化剂的作用；其次，一些 MFGM 蛋白具有广泛的生物学功能，如阻止病原体附着和参与抗菌防御活动[65-67]。因此，MFGM 成分的研究逐渐引起广泛关注。

小型泌乳动物（如牦牛和骆驼）所产的乳汁在特定地区具有重要的营养价值和经济价值[6]。除此之外，马、骆驼和牦牛乳可以作为人类饮食中的功能性食品[76]。因此，已有报道对山羊[77-78]、水牛[79]、绵羊[80]、马[81-82]和骆驼[83]这些小型泌乳动物的 MFGM 蛋白组学图谱进行了研究。此外，有研究采用 2-DE 结合质谱对两种牛科动物（契安尼那牛和荷斯坦牛）的 MFGM 蛋白进行了比较分析[84-85]。除此之外，已有研究采用 SDS-PAGE 技术对山羊、绵羊、马和骆驼等物种内的主要 MFGM 蛋白进行了比较分析[86]。山羊乳、牛乳和人乳 MFGM 中的主要蛋白通过 2-DE 结合质谱技术进行了鉴定[87]。尽管先前已存在对 MFGM 蛋白的研究，但非牛源性 MFGM 蛋白的研究相对较少且不全面。因此，对 MFGM 蛋白组成和功能的全面了解受到了一定制约。但是，已有研究表明对于小型泌乳动物（尤其是驴、牦牛和骆驼）乳需求不断增加[76]。为了更好地了解 MFGM 蛋

白的生物学功能和潜在的营养特性，需要对小型泌乳动物 MFGM 蛋白进行进一步的深入分析研究。

定量蛋白质组学方法已广泛有效地应用于生物样品中蛋白成分的同步定量和定性研究。Lable-Free 定量蛋白质组学方法已被认定为是一种可靠和多效技术[179]，广泛应用于复杂样品中特定蛋白的相对表达量的测定[69,180-181]。因此，本研究采用无标记定量蛋白质组学技术对中国荷斯坦奶牛、娟姗牛、山羊、水牛、牦牛和骆驼 MFGM 蛋白进行了分析，进而建立了上述物种 MFGM 的定量蛋白质图谱。

6.2　试验材料与方法

6.2.1　样品采集

于 2013 年 2 月至 4 月，从北京某养殖场采集奶牛（105 d±15 d）乳样 20 份，从河北某养殖场采集山羊（120 d±21 d）乳样 20 份，从河北某养殖场采集娟姗牛（107 d±20 d）乳样 20 份，从云南某养殖场采集水牛（115 d±20 d）乳样 20 份，从青海某养殖场采集牦牛（110 d±12 d）乳样 20 份，从新疆某养殖场采集马(110 d±16 d）乳样和骆驼（100 d±22 d）乳样各 8 份，乳样于-20℃条件下保存，并利用装有干冰的泡沫箱运送至中国农业科学院北京畜牧兽医研究所，于-80℃条件下保存至分析。所有供体试验动物均处于健康状态，均处于泌乳日龄约为 4 个月。

6.2.2　样品前处理

奶牛乳、水牛乳、牦牛乳、娟姗牛乳、山羊乳、骆驼乳和马乳被分离为三个部分，每个部分的处理过程如下。

（1）全乳于 4℃条件下 3 000×g 离心 15 min，取上层的乳脂。

（2）用 PBS 洗液（1L 水中含 1.91 g KH_2PO_4、15.4 g $Na_2HPO_4 \cdot 12H_2O$、80.6 g NaCl 和 2.0 g KCl）和收集的乳脂进行混合，于 39℃、30 min 的水浴后于

4 000 g、4℃离心 40 min，重复洗涤 3 次，最后用蒸馏水替代 PBS 于上述同样条件进行 1 次洗涤，即完成乳脂洗涤。

（3）向洗涤好的乳脂中加入相同体积的 4% SDS/100 mM tris-HCl 的 pH 值 7.4 的溶液于 95℃进行水浴 5 min，反复振荡、水浴 3 次以上，然后于10 000 g、4℃离心 20 min，收集下层溶液即为所需样品。

6.2.3 蛋白消解

200 μL 的蛋白质混合物和 1 mL 的丙酮混合，混合物在-20℃中保存 20 h 后 14 000 g 离心 40 min。60 μg 的蛋白质混合物被烷基化和分解。加入 100 mM 的二硫苏糖醇进行降解，95℃培养 5 min。

然后把每一份样品冷却到室温，用 200 μL UT 缓冲剂（8 M urea 和 150 mM Tris-HCl，pH 值 8.0）混合，装到过滤器（10-kDa cutoff，Sartorius，Germany）上，14 000 g 离心 15 min，再用 UT 缓冲剂洗涤一遍。接着在室温黑暗的条件下培养 30 min 后加入 100 μL 碘乙酰胺溶液（50 mM 碘乙酰胺溶于 UT 缓冲液）进行烷化，14 000 g 离心 10 min。用 100 μL UT 缓冲剂执行两次洗涤步骤，每次洗涤之后 14 000 g 离心 10 min。最后，加入 40 μL 胰蛋白酶缓冲剂（2 μg 胰蛋白酶溶于 40 μL 缓冲液），37℃培养 16~18 h。把过滤装置转移到新的试管里，14 000 g 离心 10 min，过滤收集被分解的肽，用分光光度计（Nanodrop，2000）在 OD_{280} 分析肽浓度。接着采用 C18 固相萃取柱脱盐，在真空离心蒸发浓缩器中烘干，储存在 -80℃。试验重复三次。

6.2.4 液相色谱法和串联质谱法分析

用 Thermo Fisher EASY-nLC 1000 系统加上 Q-Exactive 质谱仪分离来鉴定肽。把干燥的肽溶入缓冲剂 A（0.1%（v/v）甲酸溶于去离子水中，用 95%（v/v）缓冲剂 A 使其平衡 20 min。将 2 μg 的肽混合物放置在带有自动进样器的柱子（20 mm×100 μm，5 μm），并采用装有缓冲剂 B（84%（v/v）乙腈和 0.1%(v/v)甲酸溶于去离子水）的反相柱子（100 mm×75 μm，3 μm）中进行洗脱，流速为 400 nL/min。肽洗脱流程如下：用 0%~45%（v/v）缓冲剂 B 洗脱 100 min，45%~

100%（v/v）缓冲剂 B 洗脱 8 min，最后 100%（v/v）缓冲剂 B 洗脱 12 min。

6.2.5 蛋白质的鉴别和定量

数据源文件采用 Maxquant（version 1.3.0.5）[182] 软件分析，峰值列表是由内部的偶蹄目数据库（09-2013，108160 项）产生的。搜索参数设置如下：单一同位素集合，MS/MS 公差是±20 ppm，胰蛋白酶最多可被允许裂解两次，肽的是 2+、3+和 4+。半胱氨酸的氨基甲酸的脱甲基作为定点修改；氨基酸的氧化和蛋白质 N 端的乙酰化规定为变量修改；假定数据库模式设定在靶数据库的反相。所有报告的数据是基于 99%的置信蛋白质和多肽来鉴定，而这个鉴定结果由错误发现率（FDR）来测定，采用 2*N（诱饵）/［（N（诱饵）+N（目标）］的公式来计算 FDR，其中诱饵是反相数据库而靶子是靶子数据库。匹配在中间的流量选项设定为 2 min，识别至少两个独一无二的肽来鉴定蛋白质。此外，用 Maxquant 软件执行基础强度绝对量化选项。

6.2.6 数据分析

采用单向方差（ANOVA）分析量化蛋白，$P<0.05$ 为差异显著。采用基因本体论（GO）（http：//david.abcc.ncifcrf.gov/home.jsp.）注释分析不同物种的 MFGM 功能路径，采用 Cluster3.0 软件进行聚类分析。

6.3 结果

6.3.1 鉴定蛋白质的分析

结果显示，使用肽数据库推测出奶牛乳、水牛乳、娟姗牛乳、牦牛乳、山羊乳和骆驼乳中 MFGM 蛋白分别是 460、330、493、492、517 和 448。此外，为了确保每个物种 MFGM 中至少有两个有效值的蛋白被鉴定出来，采用 Maxquant 软件里用 iBAQ 选项将数据进行绝对量化（表 6-1）。通过 Label-Free 定量蛋白质学方法在上述几种物种中共鉴定出 653 种 MFGM。在奶牛乳、娟姗牛乳、牦牛乳和水牛乳中 MFGM 蛋白主要为乳凝集素、糖基化家族细胞黏膜分子 1、嗜乳脂蛋白亚科 1 部件 A1；在山羊乳中，MFGM 主要为组蛋白 H4、血清淀粉样蛋白 A、乳凝集素、糖基化家族细胞黏膜分子 1 和嗜乳脂蛋白亚科 1 部件 A1；在骆驼乳中，MFGM 主要为嗜乳脂蛋白亚科 1 部件 A1 和 M 片段诱导物磷酸酶 3。该结果为进一步了解不同物种乳量化蛋白质表达谱和 MFGM 的氨基酸构成提供了新的数据。

表 6-1 鉴定的奶牛乳、水牛乳、牦牛乳、娟姗牛乳、山羊乳和骆驼乳脂球膜蛋白个数

Table 6-1 Number of proteins qualitatively identified in milk fat globule membrane (MFGM) fractions from cow, buffalo, Jersey cattle, yak, goat and camel milk

	奶牛 Cow	水牛 Buffalo	娟姗牛 Jersey	牦牛 Yak	山羊 Goat	骆驼 Camel
奶牛 Cow	460	295	439	427	397	329
水牛 Buffalo		330	297	299	297	255
娟姗牛 Jersey cattle			493	444	413	351
牦牛 Yak				492	414	344
山羊 Goat					517	374
骆驼 Camel						448

6. 3. 2 鉴定蛋白质的功能分析

不同物种 MFGM 蛋白采用 GO 方法进行进一步基于其生物通路和分子功能进行分类。奶牛乳、水牛乳、娟姗牛乳、牦牛乳、山羊乳和骆驼乳通过 GO 方法分别识别出 391、285、420、412、441 和 393 种 MFGM 蛋白。其生物通路和分子功能见图 6-1和 6-2。奶牛乳、水牛乳、娟姗牛乳和牦牛乳中 MFGM 最主要的生化通路是生物调节，建立定位和定位；然而在山羊乳中 MFGM 最主要的生物过程是细胞过程，生物调节；骆驼乳 MFGM 最主要的生化过程是细胞调节。因此，生物调节、建立定位、定位和对刺激的应答是来自这些物种 MFGM 蛋白部分共同的功能。另外，在奶牛乳、水牛乳和娟姗牛乳的一些 MFGM 蛋白涉及生物附着，而水牛乳和骆驼乳 MFGM 的一些蛋白涉及细胞凋亡。而这些物种 MFGM 中最重要的分子功能是分子活性，包括蛋白合成、脂质合成、离子束缚、模式合

成、细胞表面合成、核苷酸合成及 GTP 合成，另一个主要的功能类别是酶的调节和抗氧化活性。

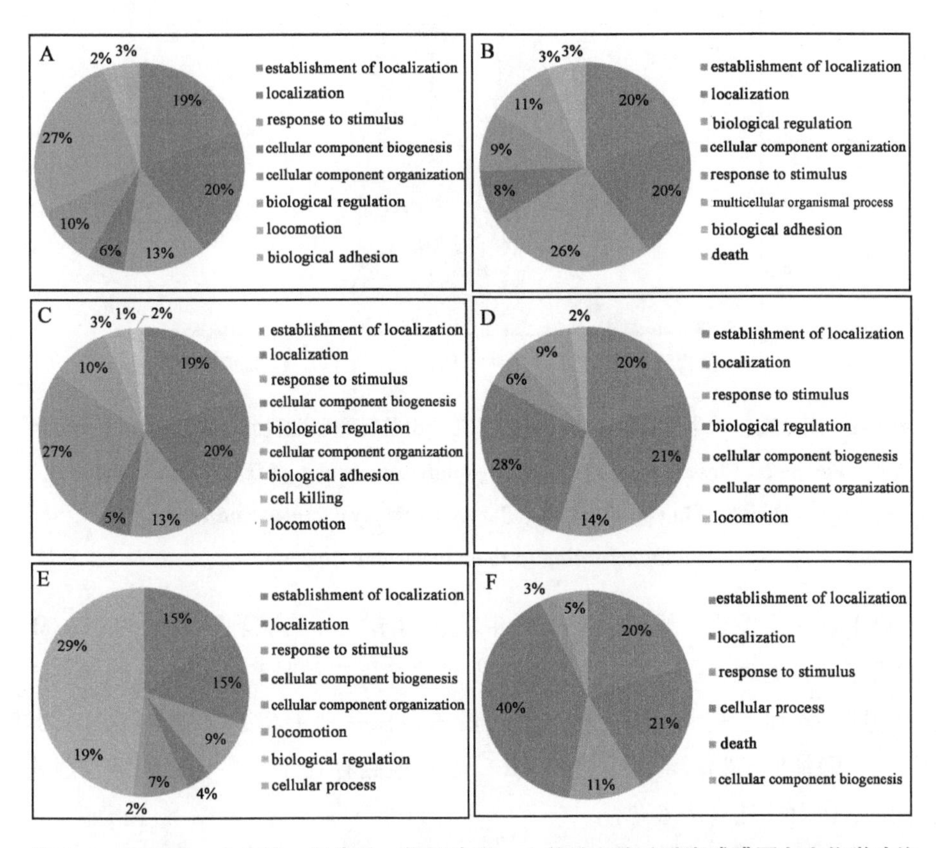

图 6-1 奶牛乳、水牛乳、牦牛乳、娟姗牛乳、山羊乳和骆驼乳脂球膜蛋白生物学功能

Fig. 6-1 Classification of milk fat globule membrane (MFGM) proteins identified in cow, buffalo, Jersey cattle, yak, goat, and camel milk based on their associated biological processes: 309 identified proteins in cow milk (A), 227 identified proteins in buffalo milk (B), 332 identified proteins in Jersey cattle milk (C), 326 identified proteins in yak milk (D), 349 identified proteins in goat milk (E), and 317 identified proteins in camel milk (F) were grouped based on their associated biological processes

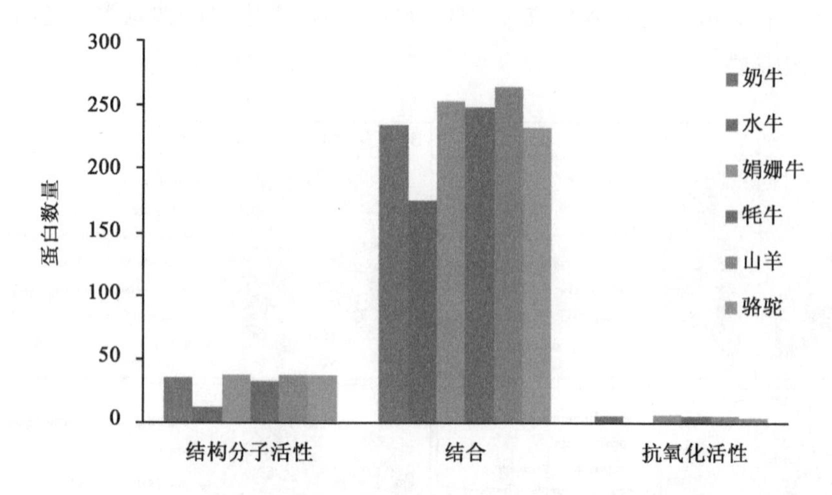

图 6-2 奶牛乳、水牛乳、牦牛乳、娟姗牛乳、山羊乳和骆驼乳脂球膜蛋白的分子功能

Fig. 6-2 Classification of milk fat globule membrane (MFGM) proteins identified in cow, buffalo, Jersey cattle, yak, goat and camel milk according to their molecular function.

奶牛乳、水牛乳、娟姗牛乳、牦牛乳、山羊乳和骆驼乳被识别的 MFGM 蛋白的通路分析见表 6-2，其中核糖体、神经营养蛋白信号和抗原加工等是共同的生物通路。然而，有个别物种的 MFGM 蛋白存在有别于其他物种的路径。例如，过氧化物酶体增殖物激活受体（PPAR）信号通路存在于水牛乳，硬结点通路存在于水牛和山羊乳（表 6-2）。

6.3.3 鉴定蛋白质的统计分析

研究各种识别的 MFGM 蛋白，共识别和量化出 215 种蛋白质，方差分析结果表明这些物种 MFGM 蛋白中的 165 种蛋白表达存在差异。其中，在这些物种中最少的也有 438 种蛋白被鉴定出来，而 72 种特定蛋白被鉴定出来。例如，封闭蛋白-1、丝氨酸肽酶抑制剂、Kunitz 型 1 仅存在于水牛乳中，MAP28 蛋白、髓过氧化物酶和二聚糖仅存在于山羊乳中，zeta 晶状体、铁蛋白、乳清酸性蛋白仅存在于骆驼乳中。

表 6-2 奶牛乳、水牛乳、牦牛乳、娟姗牛乳、山羊乳和骆驼乳脂球膜蛋白通路分析

Table 6-2 Pathway analysis of milk fat globule membrane (MFGM) proteins in cow, buffalo, Jersey cattle, yak, goat, and camel milk

物种 Species	通路 Pathway name	得分 Count	个数 its	百分含量 Percent (%)	*P* 值 *P* Value	富集 Fold enrichment
奶牛 Cow	核糖体 Ribosome	26	84	6. 65	3. 08E-17	8. 59
	膜泡运输 SNARE interactions in vesicular transport	8	38	2. 05	3. 31E-04	5. 84
	神经营养因子信号通路 Neurotrophin signaling pathway	13	118	3. 32	9. 55E-04	3. 06
	补体及凝血级联反应 Complement and coagulation cascades	9	71	2. 30	0. 0036	3. 52
	趋化因子信号转导通路 Chemokine signaling pathway	14	171	3. 58	0. 0079	2. 27
	卵母细胞减数分裂 Oocyte meiosis	10	109	2. 56	0. 0156	2. 55
	白细胞跨内皮迁移 Leukocyte transendothelial migration	10	112	2. 56	0. 0183	2. 48
	内吞作用 Endocytosis	13	179	3. 32	0. 0261	2. 02
	抗原加工和呈递 Antigen processing and presentation	7	67	1. 79	0. 0319	2. 90
	肌动蛋白细胞骨架的调控 Regulation of actin cytoskeleton	13	190	3. 32	0. 0389	1. 90
	糖酵解/糖异生 Glycolysis / Gluconeogenesis	6	55	1. 53	0. 0458	3. 03

（续表）

物种 Species	通路 Pathway name	得分 Count	个数 its	百分含量 Percent (%)	P值 P Value	富集 Fold enrichment
水牛 Buffalo	膜泡运输 SNARE interactions in vesicular transport	10	38	3. 51	2. 80E-07	10. 32
	神经营养因子信号通路 Neurotrophin signaling pathway	13	118	4. 56	3. 56E-05	4. 32
	补体及凝血级联反应 Chemokine signaling pathway	13	171	4. 56	0. 0012	2. 98
	白细胞跨内皮迁移 Leukocyte transendothelial migration	10	112	3. 51	0. 0019	3. 50
	核糖体 Ribosome	8	84	2. 81	0. 0051	3. 73
	抗原加工和呈递 Antigen processing and presentation	7	67	2. 46	0. 0066	4. 10
	补体及凝血级联反应 Complement and coagulation cascades	7	71	2. 46	0. 0087	3. 87
	卵母细胞减数分裂 Oocyte meiosis	8	109	2. 81	0. 0198	2. 88
	过氧化物酶体增殖物激活受体信号通路 PPAR signaling pathway	6	67	2. 11	0. 0265	3. 51
	硬结点 Tight junction	8	122	2. 81	0. 0341	2. 57
娟姗牛 Jersey cattle	核糖体 Ribosome	28	84	6. 67	5. 12E-19	8. 85
	补体及凝血级联反应 Complement and coagulation cascades	12	71	2. 86	5. 42E-05	4. 49
	膜泡运输 SNARE interactions in vesicular transport	8	38	1. 90	4. 34E-04	5. 59
	神经营养因子信号通路 Neurotrophin signaling pathway	13	118	3. 10	0. 0014	2. 93
	内吞作用 Endocytosis	14	179	3. 33	0. 0160	2. 08
	卵母细胞减数分裂 Oocyte meiosis	10	109	2. 38	0. 0202	2. 44
	白细胞跨内皮迁移 Leukocyte transendothelial migration	10	112	2. 38	0. 0237	2. 37
	补体及凝血级联反应 Chemokine signaling pathway	13	171	3. 10	0. 0258	2. 02
	抗原加工和呈递 Antigen processing and presentation	7	67	1. 67	0. 0383	2. 77

（续表）

物种 Species	通路 Pathway name	得分 Count	个数 its	百分含量 Percent (%)	*P* 值 *P* Value	富集 Fold enrichment
牦牛 Yak	核糖体 Ribosome	24	84	5.83	5.81E-15	7.89
	补体及凝血级联反应 Complement and coagulation cascades	12	71	2.91	3.80E-05	4.66
	神经营养因子信号通路 Neurotrophin signaling pathway	14	118	3.40	2.83E-04	3.27
	膜泡运输 SNARE interactions in vesicular transport	8	38	1.94	3.43E-04	5.81
	补体及凝血级联反应 Chemokine signaling pathway	15	171	3.64	0.0032	2.42
	白细胞跨内皮迁移 Leukocyte transendothelial migration	11	112	2.67	0.0067	2.71
	肌动蛋白细胞骨架的调控 Regulation of actin cytoskeleton	14	190	3.40	0.0188	2.03
	抗原加工和呈递 Antigen processing and presentation	7	67	1.70	0.0326	2.88
	卵母细胞减数分裂 Oocyte meiosis	9	109	2.18	0.0419	2.28
	糖酵解/糖异生 Glycolysis / Gluconeogenesis	6	55	1.46	0.0468	3.01
山羊 Goat	核糖体 Ribosome	30	84	6.80	1.05E-20	9.00
	膜泡运输 SNARE interactions in vesicular transport	9	38	2.04	9.00E-05	5.97
	神经营养因子信号通路 Neurotrophin signaling pathway	14	118	3.17	6.83E-04	2.99
	白细胞跨内皮迁移 Leukocyte transendothelial migration	13	112	2.95	0.0014	2.93
	补体及凝血级联反应 Chemokine signaling pathway	15	171	3.40	0.0072	2.21
	内吞作用 Endocytosis	14	179	3.17	0.0238	1.97
	卵母细胞减数分裂 Oocyte meiosis	10	109	2.27	0.0274	2.31
	抗原加工和呈递 Antigen processing and presentation	7	67	1.59	0.0476	2.63
	硬结点 Tight junction	10	122	2.27	0.0508	2.07

（续表）

物种 Species	通路 Pathway name	得分 Count	个数 its	百分含量 Percent (%)	*P* 值 *P* Value	富集 Fold enrichment
骆驼 Camel	核糖体 Ribosome	33	84	8.40	4.63E-26	11.15
	膜泡运输 SNARE interactions in vesicular transport	9	38	2.29	3.83E-05	6.72
	神经营养因子信号通路 Neurotrophin signaling pathway	14	118	3.56	2.13E-04	3.37
	补体及凝血级联反应 Complement and coagulation cascades	8	71	2.04	0.011524	3.20
	卵母细胞减数分裂 Oocyte meiosis	10	109	2.54	0.013577	2.60
	白细胞跨内皮迁移 Leukocyte transendothelial migration	10	112	2.54	0.016007	2.53
	补体及凝血级联反应 Chemokine signaling pathway	13	171	3.31	0.016095	2.16
	抗原加工和呈递 Antigen processing and presentation	7	67	1.78	0.028903	2.97

6.3.4 聚类分析

用Cluster 3.0软件对以上物种的MFGM蛋白进行进一步的量化分级群聚。结果显示，奶牛、娟姗牛和牦牛MFGM蛋白部分拥有相似的蛋白质组学模式而聚为一类。其中水牛隶属于这个子集并单独形成一个更大的集群，山羊乳也聚在这一类，但又独立聚一小类；而骆驼乳则为另一大类，具有一定的种属特异性（图6-3）。

6.4 讨论

在本试验中，采用Label-Free定量蛋白质组学技术用于检测奶牛乳、水牛乳、牦牛乳、娟姗牛乳、山羊乳和骆驼乳中的MFGM蛋白。在这些MFGM蛋白中共有653个蛋白点被鉴定。该结果提供了一种新的检测MFGM蛋白量化的蛋白

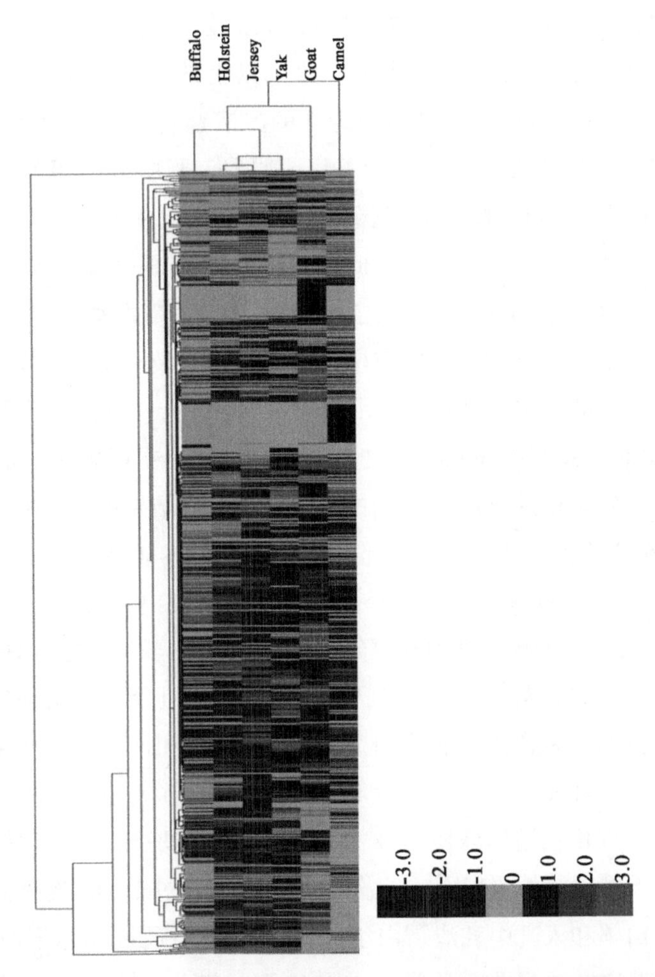

图 6-3　奶牛乳、水牛乳、牦牛乳、娟姗牛乳、山羊乳和骆驼乳脂球膜蛋白的聚类分析

Fig. 6-3　Hierarchical clustering of milk fat globule membrane（MFGM）proteins quantified in cow，buffalo，Jersey，yak，goat and camel milk. Bar color represents a logarithmic scale from −3. 0 to 3. 0

质组学方法，并揭示了不同畜种乳中 MFGM 蛋白表达的差异性。

在过去的一些研究中，已有部分有关动物 MFGM 蛋白蛋白组学方面的研究

报道。有研究通过 SDS-PAGE 结合液相质谱技术分析鉴定了奶牛 MFGM 蛋白，有 120 种蛋白被鉴定出来，且这 120 种蛋白大部分与膜转运及细胞信号传导有关，鉴定这 120 种蛋白均是基于两个或两个以上的独立的多肽序列进行确定的[71]。而另一个研究采用凝胶蛋白质组学的方法鉴定了奶牛 MFGM 主要蛋白，包括嗜乳脂蛋白家族成员-1、黄嘌呤氧化还原酶、脂肪分化相关蛋白和乳凝集素[183]。最近的研究分析了浓缩乳清蛋白和脱脂乳蛋白中的 MFGM，并分别鉴定出 244 种和 133 种 MFGM 蛋白[25]。此外，亦有研究通过滤纸收集样品再采用 nanoLC 和 LTQ Orbitrap 质谱技术，并有 169 种 MFGM 蛋白被鉴定出来[75]。而另有研究报道通过 2-DE 结合 MALDI-TOF MS 质谱技术及一维凝胶电泳集合 mLC-ESI-IT-MS-MS 质谱技术分离鉴定了水牛 MFGM 蛋白，并有 50 种蛋白被鉴定出来[79]。本试验的研究结果显示，与之前的研究相比，本试验所鉴定的奶牛、水牛、娟姗牛和牦牛乳中的 MFGM 蛋白的种类较多。该结果扩展了牛科类乳中 MF-GM 蛋白的组分类别并呈现了一个量化后的牛科类 MFGM 蛋白组分表达模式。例如，在牛科类 MFGM 蛋白中，大部分被鉴定出来的蛋白主要为嗜乳脂蛋白家族成员-1- A1、糖基化依赖性细胞黏附分子-1 和乳凝集素，与之前的报道相一致。

对于非反刍动物，如骆驼，其乳中 MFGM 蛋白组分的相关研究报道较少。有学者通过一位凝胶蛋白组学技术研究的结果指出，山羊乳中 MFGM 组分中主要为嗜乳脂蛋白、乳铁蛋白、黄嘌呤氧化酶和脂素[77-78]。最近的一项研究通过 SDS-PAGE 结合液相色谱串联质谱技术分离鉴定了骆驼乳 MFGM 蛋白，并鉴定出 322 种具有不同功能的蛋白[83]。此外，也有研究通过 2-DE 蛋白组学技术分离和鉴定奶牛、山羊和人乳中乳清蛋白组分[53]。本试验通过无标记的定量蛋白质组学技术分离鉴定非牛科类动物乳中 MFGM 蛋白组分和含量，结果显示，骆驼乳中 MFGM 蛋白组分中主要是嗜乳脂蛋白家族成员-1-A1 和 M-期诱导的磷酸酶-3，有别于牛科类 MFGM 组分。本试验的数据结果为进一步研究量化非牛科类乳 MFGM 蛋白组分提供更多的信息，且进一步描述了牛科类与非牛科类 MFGM 蛋白组分的差异性。总之，本试验结果为更深入地了解和量化哺乳类动物 MFGM 组分提供了一定的理论基础。

MFGM 蛋白含有大量不同种类的蛋白。在乳脂球分泌的过程中，甘油三酯通

过复杂的脂质双分子层膜形成一个通道进而在乳腺上皮细胞中形成乳脂，而细胞质合成所需的组分保留在不同的膜层间[60]。本试验中，不同物种乳中 MFGM 均含有酪蛋白和主要的乳清蛋白，这与前人的研究结果显示的奶牛、山羊和骆驼 MFGM 蛋白组学的研究中也同样检测出酪蛋白和主要的乳清蛋白的结果相一致[25,75,83,87]。此外，有研究通过无标记的定量蛋白质组学技术在人乳 MFGM 中鉴定出乳中的 3 种酪蛋白，包括 α-S_1酪蛋白、β-酪蛋白和 κ-酪蛋白以及不同的乳清蛋白，包括 α-乳清蛋白和乳铁蛋白[69]。这些结果表明，酪蛋白和主要的乳清蛋白均是 MFGM 的组分，并且参与乳腺上皮细胞分泌乳脂球的代谢机制过程。

MFGM 蛋白与许多生物学功能的发挥相关联。前人的研究结果显示，被鉴定出的奶牛乳中 MFGM 蛋白中有 23%与膜蛋白的转运相关，23%与细胞信号传导有关[71]。另一项研究通过利用 Swiss-Prot 数据库对奶牛乳 MFGM 蛋白进行归类并指出牛乳 MFGM 中最重要的一类蛋白与机体免疫与防御机制有关[75]。最近的一项研究结果显示，骆驼乳 MFGM 蛋白主要参与蛋白质转运、脂质合成和肌动蛋白细胞骨架构建，且这些结果是通过利用数据库进行注释、比对以及综合生物信息学资源发现所得[83]。然而，由于被鉴定的其他小物种乳 MFGM 蛋白太少而不能利用数据库注释的信息进行生物学功能分析。本试验的结果显示，大量的 MFGM 蛋白被鉴定出来，为进一步深入研究不同物种乳 MFGM 提供新的依据，同时，为进一步揭示 MFGM 蛋白的生物学功能和潜在的营养作用提供一定的信息。此外，本试验中所有物种乳 MFGM 蛋白的最常见的功能主要为生物功能调节、应激响应等。几种通路，包括核糖体、神经调节以及抗原形成与呈现等是不同物种乳中最常见的几种路径。此外，本研究结果通过聚类分析奶牛乳、水牛乳、牦牛乳、娟姗牛乳、山羊乳和骆驼乳 MFGM 蛋白的结果与之前通过同位素标记相对和绝对定量测定的奶牛乳、水牛乳、牦牛乳、山羊乳和骆驼乳清蛋白进行的聚类分析的结果是一致的，都将以上物种分为三类[53]。综合以上研究结果，揭示了不同物种乳蛋白的内在特征性。

6.5 小结

（1）通过无标记的定量蛋白质组学技术分离鉴定奶牛乳、水牛乳、牦牛乳、娟姗牛乳、山羊乳和骆驼乳中的 MFGM 蛋白，并有 653 种蛋白被鉴定出来。

（2）通过 GO 分析注释，确定这些蛋白主要生物学功能和相关的调节路径。

（3）通过 Cluster 聚类分析奶牛、水牛、牦牛、娟姗牛、山羊和骆驼乳中的 MFGM 蛋白的结果得出，以上物种主要聚为三大类，其中四种牛科类：奶牛乳、水牛乳、牦牛乳和娟姗牛乳聚为一类，其中水牛隶属于这个子集并单独形成一个更大的集群，山羊乳也聚在这一类，但又独立聚一小类；而骆驼乳则为另一大类，具有一定的种属特异性。

7 基于转录组学方法分析新鲜奶牛与新鲜山羊乳清中特征性microRNA及反复冻融奶牛乳清与新鲜奶牛乳清样microRNA的差异性

7.1 引言

过去10年内，microRNA越来越明确的被证明是小的非编码RNA的一大类，其通过调节基因的表达参与广泛的细胞活动过程。经过10多年的研究，人类已经发现哺乳动物中microRNA的存在，并探索了它们在疾病治疗方面的作用。关于microRNA领域的快速发展，越来越需要适用的功能验证方法来检验microRNA的表达和生物学功能。MicroRNA在2000年以前都仅限于在非哺乳或非脊椎动物中的研究。

乳是哺乳动物新生儿的唯一营养来源，但是为了适合子代发育，不同的哺乳动物乳的成分也不尽相同[147-148]。Kosaka等（2010b）报道指出，人类母乳中含有免疫相关的microRNA，但不含器官特异性的miRNA（血细胞、肌肉、胰腺和肝脏特异性）[149]。有研究报道人乳中存在microRNA[152]，最近其他研究者还报道了牛乳中也存在miRNA。Izumi等（2012）使用基因芯片和定量PCR研究牛乳的miRNA，验证初乳和常乳之间的差异，初乳乳清中RNA的浓度高于常乳乳清RNA浓度。牛奶中共检测出102种microRNA，与免疫和发育相关的microRNA包括miR-15b、miR-27b、miR-34a、miR-106b、miR-130a、miR-155和miR-223。使用定量PCR检测牛乳中的这些microRNA，以上每种microRNA在初乳中的表达都显著高于常乳中的表达，证实了牛乳中mRNA的存在。尽管如此，原料奶乳清中合成的miRNA被降解，天然存在的miRNA和mRNA可以抵抗酸性条件

和RNA酶的作用，也就是说，牛奶中的RNA分子非常稳定[151]。此外，Weber等（2010）检测了12种人类体液中的microRNA，其中包括人乳，但相比Kosaka等（2010b）的研究结果发现，不同的乳源背景，其所含有的microRNA种类和表达丰度不同[152]。因此，不同畜种乳中的microRNA是否具有一定的物种特异性，需要进一步深入研究。

基于高通量测序测定microRNA，使得更多的microRNA被发现，也扩展了microRNA的研究方向，进而帮助人们更为深入地了解microRNA。因此本试验通过采用Illumina（USA）的Hiseq2500测序平台，分析测定牛乳清和羊乳清中的SE50（单端独长50bp）的microRNA，建立奶牛乳清和山羊乳清中特征性microRNA相关数据信息，为进一步确定不同物种乳中特征性的microRNA提供研究基础，为区分不同畜种乳提供依据。

此外，在体液中检测少量的microRNA要求复杂和灵敏的提取方法，因此需要设计切实可行的检测方法。microRNA表达具有组织特异性，不存在明显的性别差异，也不存在显著的个体差异。microRNA可以在体外定量，实时定量PCR（RT-PCR）方法可以很精确地定量分析microRNA的表达，这是一种快速有效的检测方法，本试验根据高通量测序的结果，分析确定鲜牛乳清和鲜羊乳清中差异microRNA，并进一步通过实时定量PCR方法定量验证分析牛乳清和山羊乳清差异microRNA的表达。

7.2 试验材料与方法

7.2.1 样品采集

（1）于2013年4月，从北京某养殖场采集奶牛单头乳样60份，于4℃条件下冷藏保存，送至中国农业科学院北京畜牧兽医研究所，随机取30份混合为一个样本，制备2个新鲜奶牛乳混合样本（C-1，C-2）；从河北某养殖场采集山羊单头乳样50份，于4℃条件下冷藏保存，送至中国农业科学院北京畜牧兽医研究所，随机取25份混合为一个样本，制备2个新鲜山羊乳混合样本（G-1，G-

2)；同时，将相应的乳样进行反复冻融制备 2 个反复冻融奶牛乳样混合样本（CA1，CA2）和 2 个反复冻融山羊乳样混合样本（GA1，GA2），所采集乳样供体动物均处于泌乳日龄约为 4 个月；

（2）20 份鲜牛乳样和 20 份鲜羊乳样用于差异 microRNA 的 RT-PCR 检测。

7.2.2 乳清的提取

（1）将全乳于 4℃条件下 1 200×*g* 离心 10 min，去脂、去细胞及去大的杂质；

（2）将去脂后的部分于 4℃条件下21 500×*g* 离心 30 min，去除残留的脂肪和酪蛋白；

（3）再将去脂后的部分于 4℃条件下 21 500×*g* 离心 60 min，去除残留的脂肪和酪蛋白；

（4）将去脂后的乳清部分分别通过 0.65 μm、0.45 μm 和 0.22 μm 的滤膜去除细胞杂质，所得的溶液即为所需乳清样品。

7.2.3 乳清总 RNA 的提取

乳清中仍然含有大量的蛋白质，因此，我们使用硅胶模柱的 RNA 提取试剂盒纯化 RNA，采用的提取试剂盒为 miRNeasy Serum/Plasma Kit（Qiagen，Hilden，Germany）：

（1）将 1 mL（5 倍于乳清样本体积）QIAzolLysis Reagent 加入 200 μL 乳清样品中，剧烈振荡，充分混匀后室温孵育 5 min（使样品变性）。

（2）将 200 μL（体积与起始样品体积相同）氯仿加入变性样品中，剧烈振荡 15 s，室温孵育 3 min，然后在 4℃，12 000×*g* 条件下离心 15 min。

（3）所得到的水相转移到新的 2 mL 离心管中（如果乳清的体积为 200 μL，那么所得水相的近似体积为 770 μL），然后添加 1 155 μL（水相体积的 1.5 倍）100 %分子生物学级乙醇，上下颠倒混匀。

（4）从上一步样品中吸取 700 μL 加入柱内，≥8 000×*g*（≥10 000 r/min），室温离心 15 s。

（5）重复上一步，丢弃流通物。

（6）加 700 μL Buffer RWT 到柱内，≥8 000×g（≥10 000 r/min），室温离心 15 s（洗柱），丢弃流通液。

（7）加 500 μL Buffer RPE 到柱内，≥8 000×g（≥10 000 r/min）、室温离心 15 s（洗柱），丢弃流通液。

（8）吸取 500 μL 80%的乙醇（RNase-free water 配制）到柱内，≥8 000×g（≥10 000 r/min）离心 2 min（洗膜），连同流通液丢弃柱下的收集管。

（9）将柱子放到试剂盒提供的新的 2 mL 收集管中，开盖全速离心 5 min，弃管。

（10）将柱子放入新的试剂盒提供的 1.5 mL 收集管中，在膜的中间直接加入 14 μL RNase-free water，全速离心 1 min 收集 RNA，迅速-80℃保存，并尽快上机测序。

7.2.4　高通量测序分析仪

Illumina（USA）的 Hiseq2500 测序平台。

7.2.5　建库测序实验流程

建库测序实验流程如图 7-1 所示。

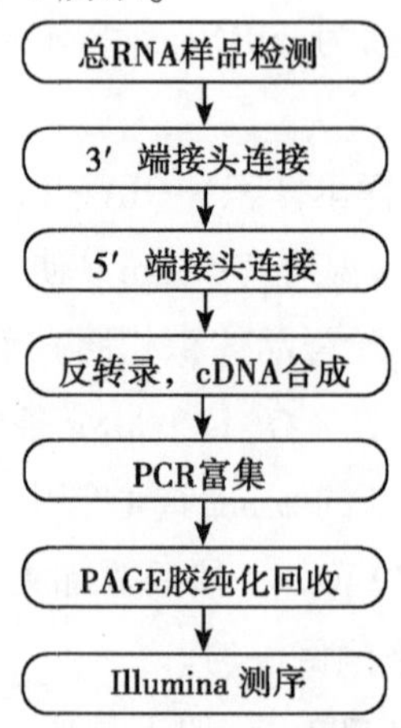

图 7-1　建库测序实验流程

Fig. 7-1　The process of building the sequencing library

7.2.5.1 总 RNA 样品检测

（1）琼脂糖凝胶电泳分析 RNA 降解程度以及是否有污染。

（2）Nanodrop 检测 RNA 的纯度（OD260/280 比值）。

（3）Qubit 对 RNA 浓度进行精确定量。

（4）Agilent 2100 精确检测 RNA 的完整性。

7.2.5.2 文库构建

样品检测合格后，使用 Small RNA Sample Pre Kit 构建文库，利用 Small RNA 的 3′及 5′端特殊结构（5′端有完整的磷酸基团，3′端有羟基），以总 RNA 为起始样品，直接将小 RNA 两端加上接头，然后反转录合成 cDNA。随后经过 PCR 扩增，PAGE 胶电泳分离目标 DNA 片段，切胶回收得到的即为 cDNA 文库。构建原理如图 7-2 所示。

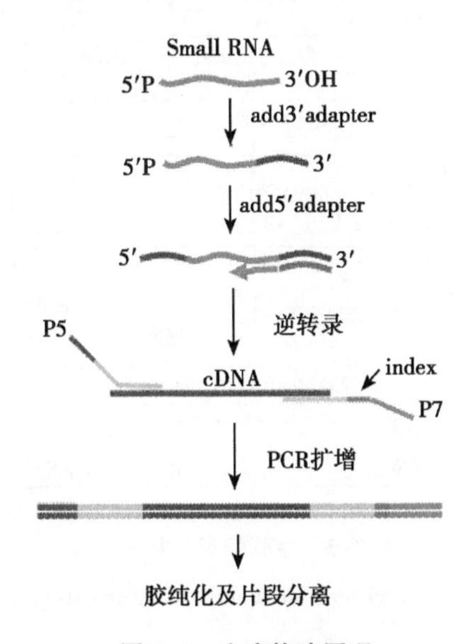

图 7-2 文库构建原理

Fig. 7-2 The principle of building the library

7.2.5.3 库检

文库构建完成后，先使用 Qubit2. 0 进行初步定量，稀释文库至 1 ng/uL，随后使用 Agilent 2100 对文库的插入片段进行检测，插入片段符合预期后，使用 Q-PCR

方法对文库的有效浓度进行准确定量（文库有效浓度>2 nM），以保证文库质量。

7.2.5.4 上机测序

库检合格后，把不同文库按照有效浓度及目标下机数据量的需求 pooling 后进行 HiSeq/MiSeq 测序。

7.2.6 生物信息分析流程

生物信息分析流程如图 7-3 所示。

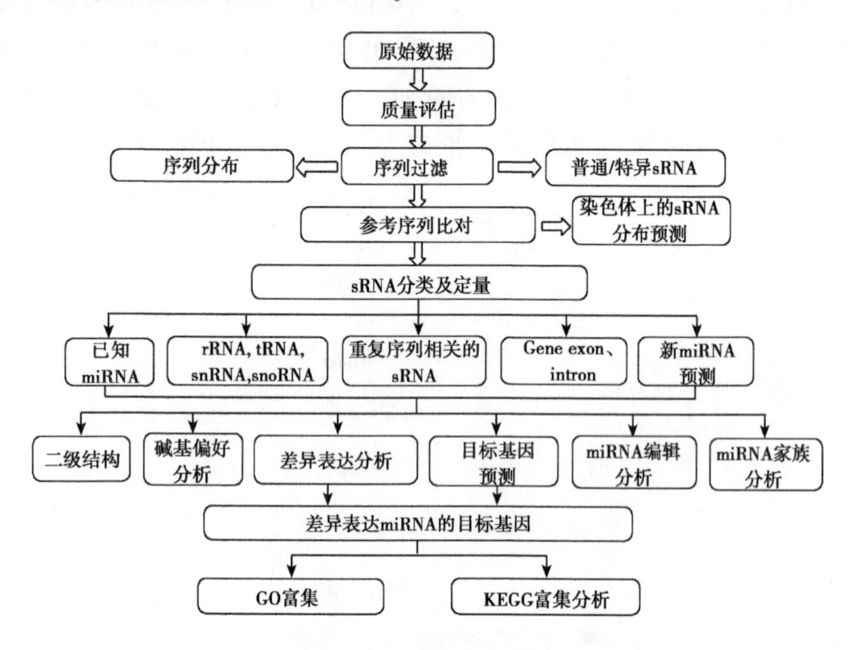

图 7-3　生物信息分析流程

Fig. 7-3　The process of biological information analysis

7.2.7 RT-PCR

7.2.7.1 乳清总 RNA 的提取

乳清中仍然含有大量的蛋白质，因此，我们使用硅胶模柱的 RNA 提取试剂盒纯化 RNA，采用的提取试剂盒为 miRNeasy Serum/Plasma Kit（Qiagen，Hilden，

Germany)，此外，采用 Serum/Plasma Spike-In Control（1.6×10^8copies/μL，Syn-cel-miR-39 miScript miRNA Mimic）作为外参：

（1）将 1 mL（5 倍于乳清样本体积）QIAzolLysis Reagent 加入 200 μL 乳清样品中，剧烈振荡，充分混匀后室温孵育 5 min（使样品变性）。

（2）将 3.5 μL miRNeasy Serum/Plasma Spike-In Control（1.6×10^8copies/μL，Syn-cel-miR-39 miScriptmiRNA Mimic）加入 1.2 mL 变性样品中（乳清+QIAzol），充分混匀。

（3）将 200 μL（体积与起始样品体积相同）氯仿加入变性样品中，剧烈振荡 15 s，室温孵育 3 min，然后在 4℃，12 000×g 条件下离心 15 min。

（4）所得到的水相转移到新的 2 mL 离心管中（如果乳清的体积为 200 μL，那么所得水相的近似体积为 770 μL），然后添加 1 155 μL（水相体积的 1.5 倍）100 %分子生物学级乙醇，上下颠倒混匀。

（5）从上一步样品中吸取 700 μL 加入柱内，≥8 000×g（≥10 000 r/min）、室温离心 15 s。

（6）重复上一步，丢弃流通物。

（7）加 700 μL Buffer RWT 到柱内，≥8 000×g（≥10 000 r/min）、室温离心 15 s（洗柱），丢弃流通液。

（8）加 500 μL Buffer RPE 到柱内，≥8 000×g（≥10 000 r/min）、室温离心 15 s（洗柱），丢弃流通液。

（9）吸取 500 μL 80%的乙醇（RNase-free water 配制）到柱内，≥8 000×g（≥10 000 r/min）离心 2 min（洗膜），连同流通液丢弃柱下的收集管。

（10）将柱子放到试剂盒提供的新的 2 mL 收集管中，开盖全速离心 5 min，弃管；将柱子放入新的试剂盒提供的 1.5 mL 收集管中，在膜的中间直接加入 14 μL RNase-free water，全速离心 1 min 收集 RNA。

7.2.7.2 逆转录反应

本试验逆转录反应使用 PrimeScript © miRNA RT-PCR Kit（TaKaRa）。

（1）反应体系

2×miRNA Reaction Buffre Mix（for Real Time）：10 μL；

0.1%BSA：2 μL；

miRNA PrimeScript RT Enzyme Mix：2 μL；

Total RNA：2 μL；

RNase-Free H_2O：4 μL；

（2）逆转录反应条件

37℃ 60 min，85℃ 5 s，PCR 仪（Applied Biosystems）设定条件进行。

7.2.7.3 PCR 反应体系

本试验 PCR 反应使用 PrimeScript® miRNA RT-PCR Kit（TaKaRa）（TaKaRa），在 ABI Prism 7 500 SDS Instrument（Applied Biosystems）上进行。反应体系及反应条件如下。

（1）荧光定量 RT-PCR 反应体系（20 μL）

SYBR Premix Ex TaqⅡ（2×）：10 μL

PCR Forward Primer（10 μM）:0.8 μL

Uni-miRqPCR Primer（10 μM）:0.8 μL

ROX Reference Dye（50×）:0.4 μL

模板（cDNA 溶液）：2 μL

dH_2O：6 μL

（2）荧光定量 RT-PCR 反应条件

步骤 1　预变性

Reps：1

95℃ 30 s

步骤 2　PCR 反应

Reps：40

95℃ 5 s

60℃ 34 s

步骤 3　融解阶段

95℃ 15 s

60℃ 60 s

95℃ 15 s

观察融解曲线和扩增曲线，Ct 数据使用7 500 Sequence Detection Systems Software（Applied Biosystems）导出和分析，靶基因表达量使用相对 Ct 法检测（Livak&Schmittgen，2001），通过这个方法，靶基因通过外参果蝇的 Syn-cel-miR-39 得以标准化。采用 $2^{-\Delta\Delta CT}$ 相对定量方法进行计算，数据表示为平均值（$\bar{x}$）±标准误（SEM）。每个基因设 3 个技术重复。

7.2.8　数据分析

7.2.8.1　miRNA 表达水平分析

对各样本中已知和新 miRNA 进行表达量的统计，并用 TPM（Zhou 等，2010）[184]进行表达量归一化处理。

7.2.8.2　miRNA 表达量 TPM 密度分布图

TPM 密度分布能整体检查样品的基因表达模式（图 7-4）。

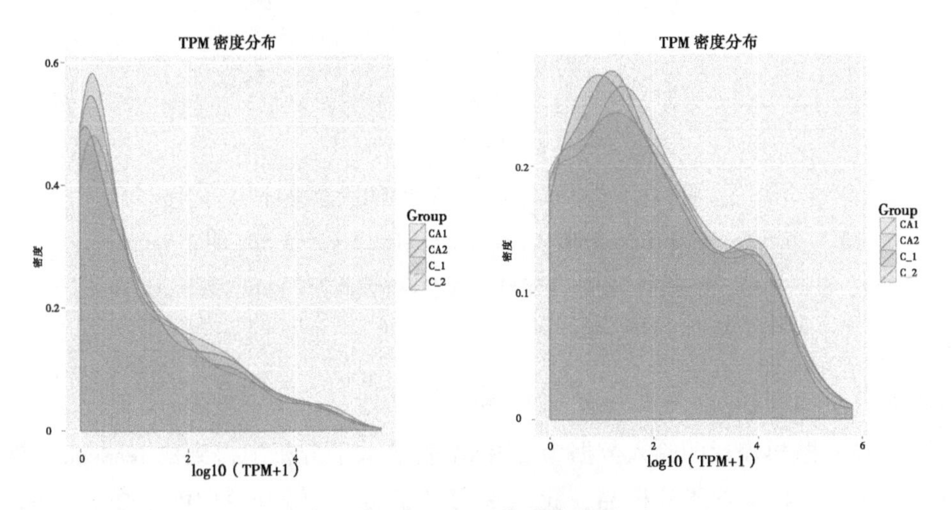

图 7-4　TPM 密度分布图

Fig. 7-4　TPM density distribution

注：横坐标为基因的 log10（TPM+1）值，纵坐标为对应 log10（TPM+1）的密度

7.2.8.3 样品间相关性检查

样品间基因表达水平相关性是检验实验可靠性和样本选择是否合理性的重要指标。相关系数越接近 1，表明样品之间表达模式的相似度越高。若样品中有生物学重复，通常生物重复间相关系数要求较高（图 7-5~图 7-8）。

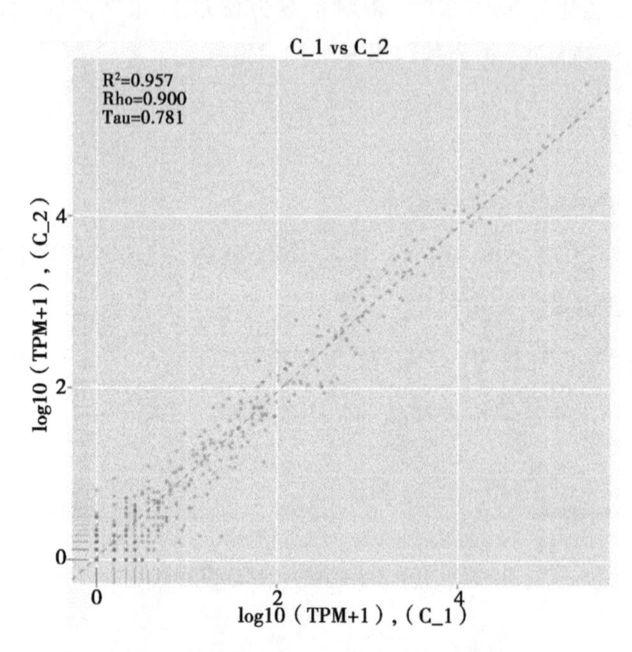

图 7-5 奶牛样品间基因表达量相关性分析

Fig. 7-5 Correlation analysis of gene expression between cow milk samples

注：横坐标为样品 1 的 log10（TPM+1），纵坐标为样品 2 的 log10（TPM+1），R^2：pearson 相关系数的平方；Rho：spearman 相关系数；Tau：kendall-tau 相关系数

7.2.8.4 miRNA 差异表达分析结果

miRNA 差异表达的输入数据为 miRNA 表达水平分析中得到的 readcount 数据。对于有生物学重复的样品，我们采用基于负二项分布的 DESeq2（Anders 等，2010）[185]进行差异分析；对于无生物学重复的样品，先采用 TMM 对 readcount 数据进行标准化处理，之后用 DEGseq（Wang 等，2010）[186]进行差异分析。

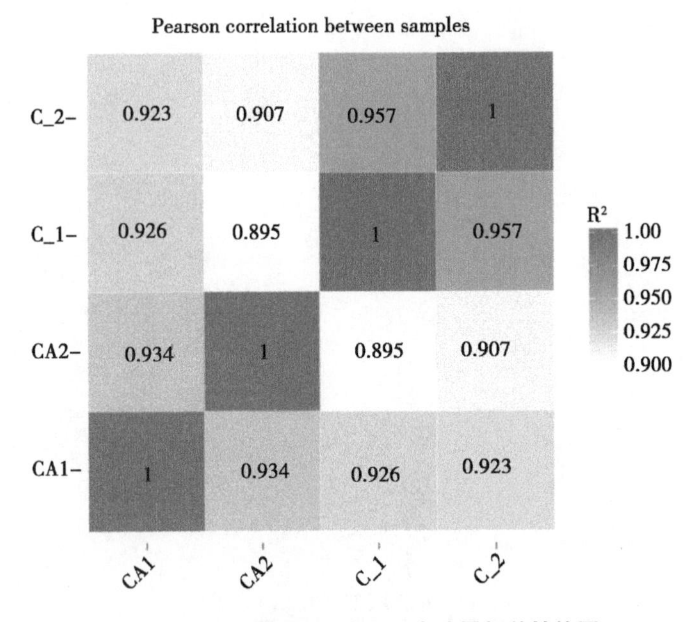

图 7-6　奶牛乳样品间 miRNA 表达量相关性热图

Fig. 7-6　Pearson correlation between cow milk samples

7.2.9　差异 miRNA 筛选

从差异倍数（Fold change）和校正后的显著水平（padj/qvalue）两个方面进行评估，对差异 miRNA 进行筛选，当样品有生物学重复时，差异 miRNA 的筛选条件为：padj<0.05。

7.3　结果

7.3.1　鲜牛奶乳清与鲜羊奶乳清中 mincroRNA 的鉴定

生鲜牛乳和山羊乳中 microRNA 的种类和含量存在显著差异。差异 microRNA 维恩图见图 7-9，总的来说，生鲜牛乳中检测到 415 种 microRNAs，其中 381 种

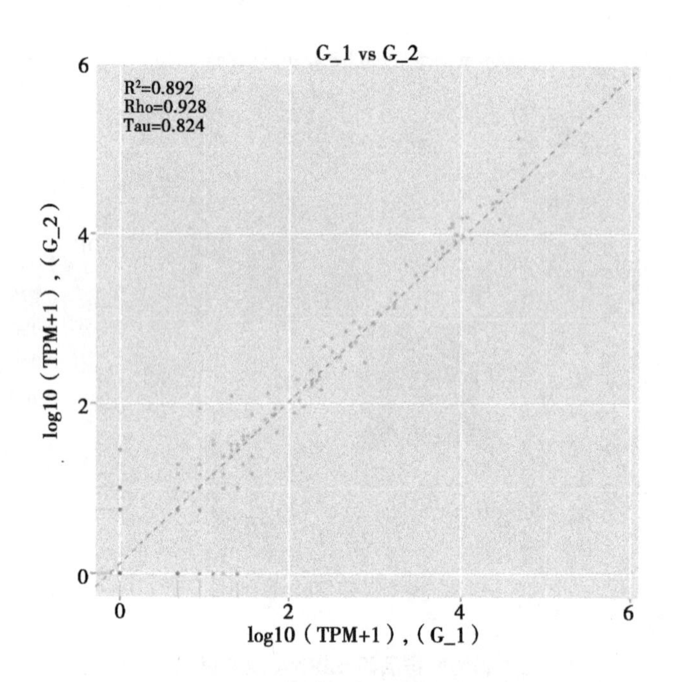

图 7-7　山羊乳样品间基因表达量相关性分析

Fig. 7-7　Correlation analysis of gene expression between goat milk samples

注：横坐标为样品 1 的 log10（TPM+1），纵坐标为样品 2 的 log10（TPM+1），R^2：pearson 相关系数的平方；Rho：spearman

microRNAs 具有典型的颈环发夹结构与已知的 microRNAs 匹配，此外，还鉴定出 34 种新的 microRNAs；在生鲜羊乳中检测到 111 种 microRNAs，其中有 13 种新的 microRNAs；而其中的 29 种 microRNAs 为鲜牛奶乳清和鲜羊奶乳清共有的。

7.3.2　鲜牛奶与冻融牛奶乳清中差异 microRNA 聚类分析

与新鲜牛乳清样本相比，反复冻融后 microRNA-1246、microRNA-135a、microRNA-15b、microRNA-223、microRNA-30b-5p、microRNA-342、microRNA-365-3p、microRNA-378b、microRNA-500、microRNA-664b 和 microRNA-885 表达上调，而 let-7a-5p、let-7c、let-7e、microRNA-122、microRNA-193a-5p、microRNA-25、microRNA-3432、microRNA-423-5p 和 microRNA-760-3p 表达下

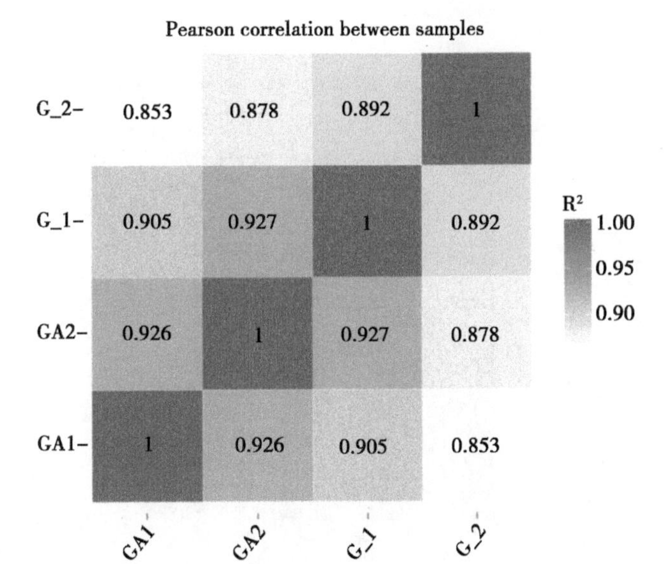

图 7-8　山羊乳样品间 miRNA 表达量相关性热图

Fig. 7-8　Pearson correlation between goat milk samples

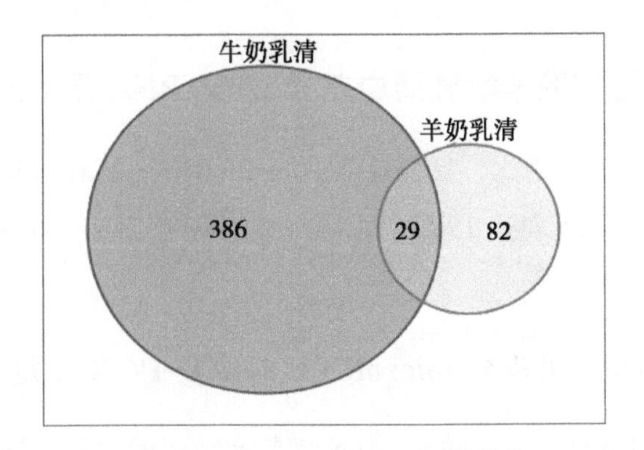

图 7-9　鲜牛奶乳清与鲜羊奶乳清中 mincroRNA 对比维恩图

Fig. 7-9　The Venn diagram of microRNA in fresh cow milk and fresh goat milk

调（图 7-10）。

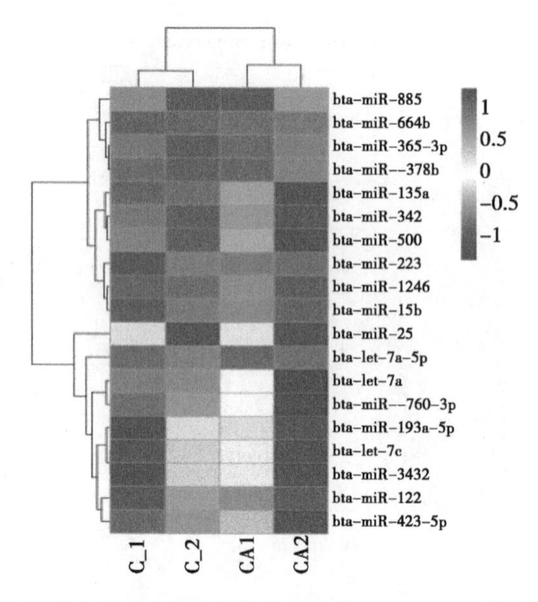

图 7-10　鲜牛奶与冻融牛奶乳清中差异 microRNA 聚类分析图

Fig. 7-10　Cluster analysis of differentially expressed microRNA in fresh cow milk and multigelation cow milk

7.3.3　鲜羊奶与冻融羊奶乳清中差异 microRNA 聚类分析

与新鲜羊乳清样本相比，反复冻融后 microRNA-23a、microRNA-23b、microRNA-30d 和 novel-7 表达上调，而 let-7g、microRNA-19b 和 microRNA-26b 表达下调（图 7-11）。

7.3.4　鲜牛奶和羊奶差异 microRNA 的 RT-PCR 测定

在体液中检测少量的 microRNA 要求复杂和敏感的提取方法，因此需要设计切实可行的检测方法。miRNA 表达具有组织特异性，不存在明显的性别差异，也不存在显著的个体差异。miRNA 可以在体外定量，实时定量 PCR 方法可以很精确地定量分析 miRNA 的表达，这是一种快速有效的检测方法。

本试验根据高通量测序所鉴定出的鲜牛奶乳清与鲜羊奶乳清中各自特征性的

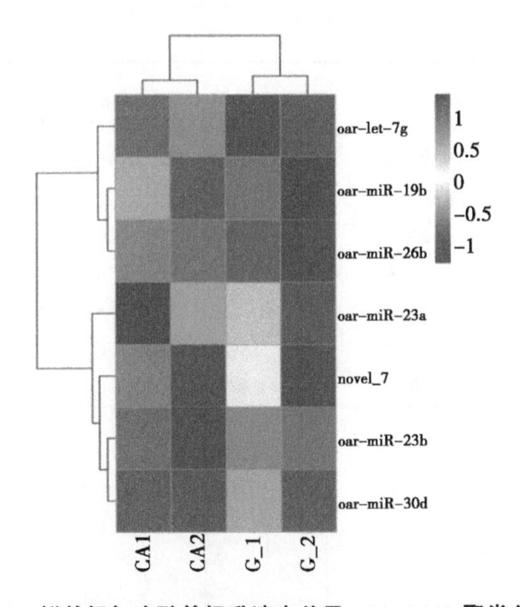

图 7-11 鲜羊奶与冻融羊奶乳清中差异 microRNA 聚类分析图

Fig. 7-11 Cluster analysis of differentially expressed microRNA in fresh cow milk and multigelation cow milk

mincroRNA 的表达量，以 padj<0.05 为检测差异 microRNA 的依据，由鲜牛乳清中 386 种特征性 microRNA、鲜羊乳清中 82 种特征性 microRNA 以及 29 种鲜牛羊乳清中共有的 microRNA 的相对表达量进行筛选，分别筛选出 4 个表达量高的microRNA（表 7-1）进行引物设计并进行 RT-PCR 定量检测（图 7-13，图 7-14）。

表 7-1 鲜牛奶乳清、鲜羊奶乳清各自特有及其共有 microRNA 名称及序列

Table 7-1 The respective characteristic and common microRNA sequence in fresh cow whey and fresh goat whey

	sRNA	序列（5′to3′）
牛奶乳清特有	bta-miR-200c	UAAUACUGCCGGGUAAUGAUGGA
	bta-let-7b	UGAGGUAGUAGGUUGUGUGGUU
	bta-miR-3596	AACCACACAACCUACUACCUCA
	bta-miR-26c	AGCCUAUCCUGGAUUACUUGAA

（续表）

	sRNA	序列（5′to3′）
羊奶乳清特有	oar-miR-30a-5p	UGUAAACAUCCUCGACUGGAAGC
	oar-miR-21	UAGCUUAUCAGACUGAUGUUGAC
	oar-miR-200c	UAAUACUGCCGGGUAAUGAUGG
	oar-miR-30d	UGUAAACAUCCCCGACUGG
牛羊乳清共有	bta-miR-148a	UCAGUGCACUACAGAACUUUGU
	bta-miR-26a	UUCAAGUAAUCCAGGAUAGGCU
	bta-miR-22-3p	AAGCUGCCAGUUGAAGAACUG
	bta-miR-200b	UAAUACUGCCUGGUAAUGAUG

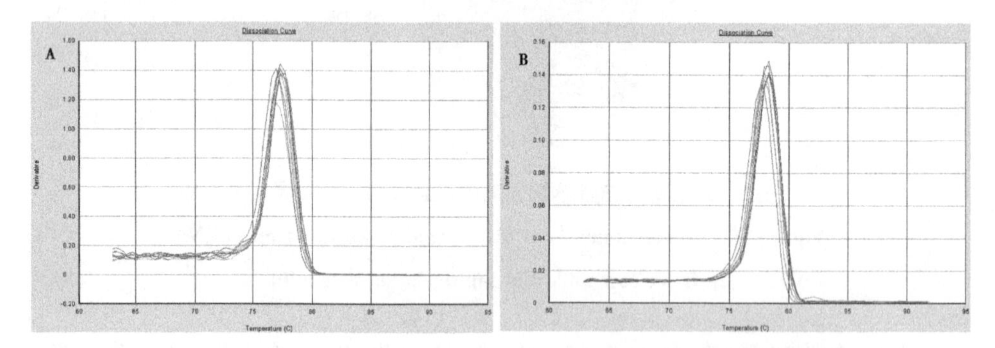

图 7-12　RT-PCR 溶解曲线（A：鲜牛奶乳清；B：鲜羊奶乳清）

Fig. 7-12　Dissociation curve by RT-PCR（A：fresh cow whey；B：fresh goat whey）

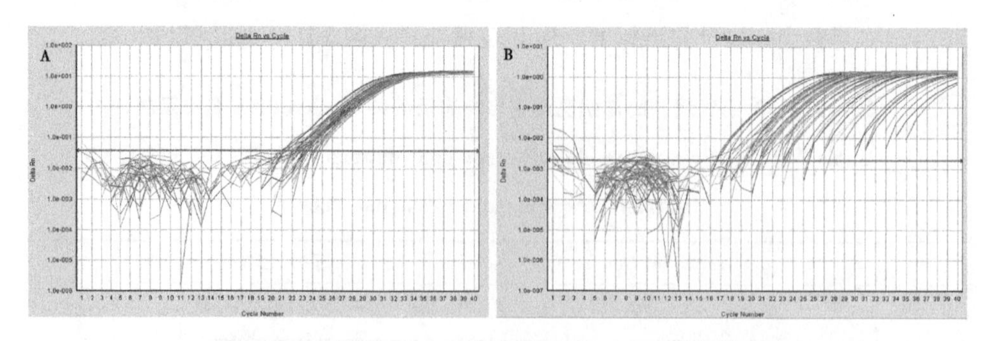

图 7-13　扩增曲线（A：鲜牛奶乳清；B：鲜羊奶乳清）

Fig. 7-13　Amplification plot by RT-PCR（A：fresh cow whey；B：fresh goat whey）

由图 7-14 的结果可以看出，通过高通量测序所筛选出的表达丰度较高的鲜牛奶乳清中特有的 microRNA（microRNA-200c、let-7b、microRNA-3596 和 microRNA-26c）均在单个鲜牛奶乳清中检测出，且以 Syn-cel-microRNA-39 为外参的基础上进行校正，表达丰度均较高，除 microRNA-26c 外，其余的均较外参 microRNA-39 的相对表达量高；其中 let-7b、microRNA-3596 和 microRNA-26c 均未在鲜羊奶乳清中检测出，呈现出较强的物种特异性，而 microRNA-200c 则

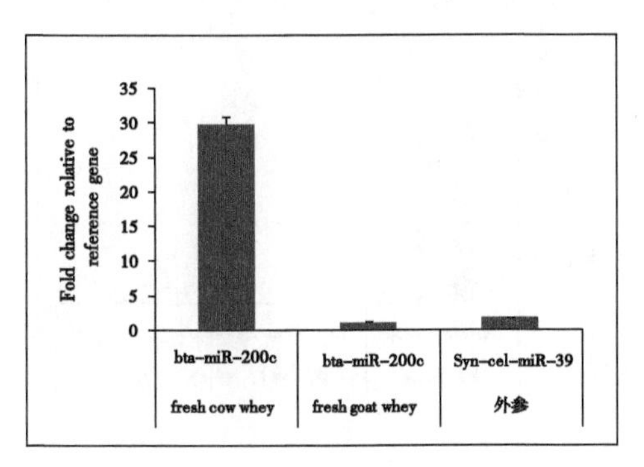

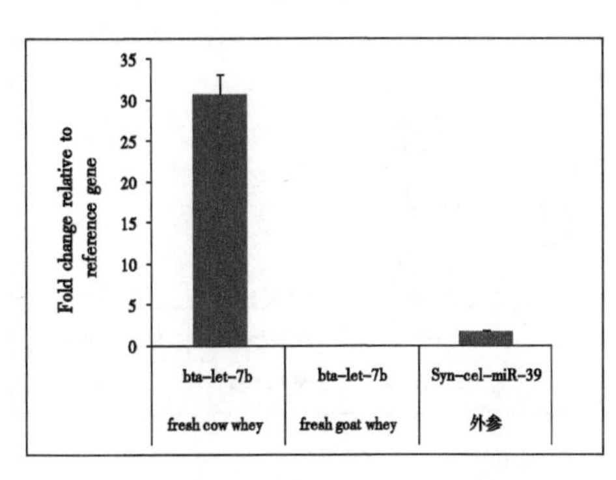

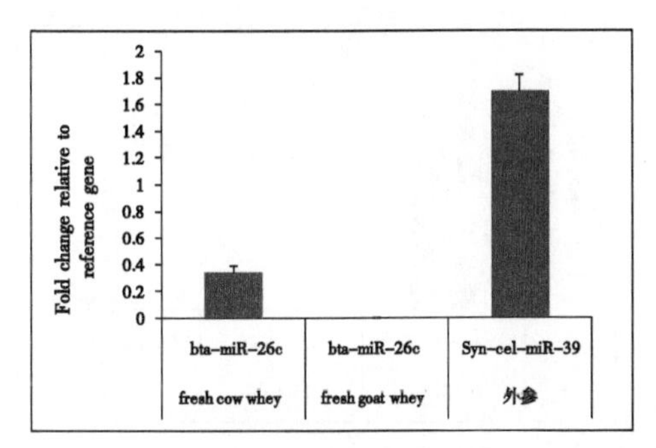

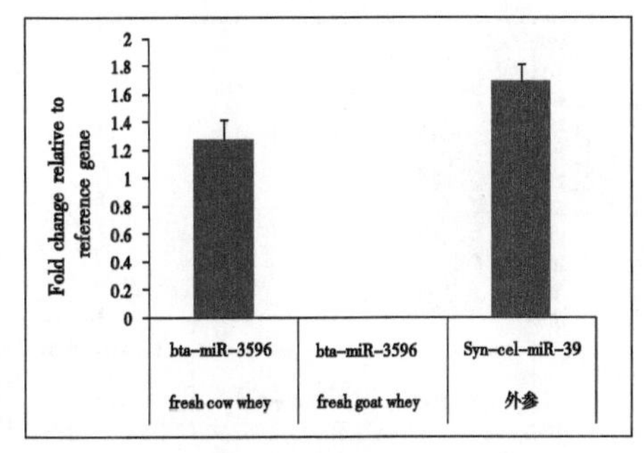

图 7-14　鲜牛奶乳清中 microRNA-200c、let-7b、microRNA-3596 和 microRNA-26c 相对表达量

Fig. 7-14　The relative expression of selected microRNA（miRNA）- microRNA-200c、let-7b、microRNA-3596 and microRNA-26c in fresh cow whey with normalization by RT-PCR

注：采用果蝇 microRNA-39 进行标准化（To measure the levels of expression of miRNA in equal volumes of whey fraction from cow whey, the quantitative PCR data obtained were normalized to spiked-in synthetic cel-miR-39）

在鲜羊奶乳清中检测出。

由图 7-15 的结果可以看出，通过高通量测序所筛选出的表达丰度较高鲜羊奶乳清中特有的 microRNA（microRNA-30a-5p、microRNA-200c、microRNA-21 和 microRNA-30d）均在单个鲜羊奶乳清中检测出，且以 Syn-cel-microRNA-39 为外参的基础上进行校正，表达丰度均较高，且其中的 microRNA-30-5p、microRNA-21 和 microRNA-30d 均未在鲜牛奶乳清中检测出，呈现出较强的物种特

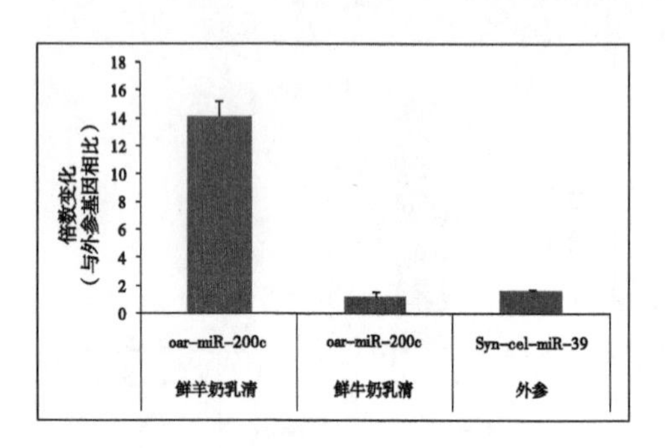

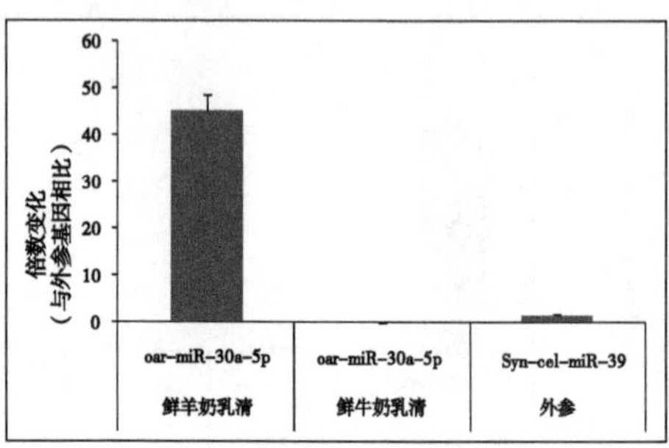

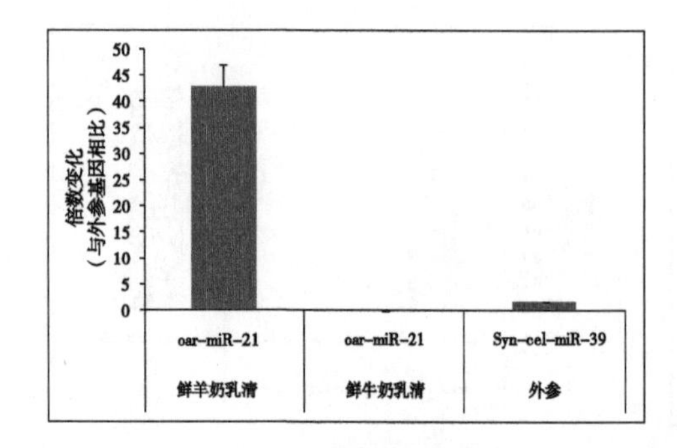

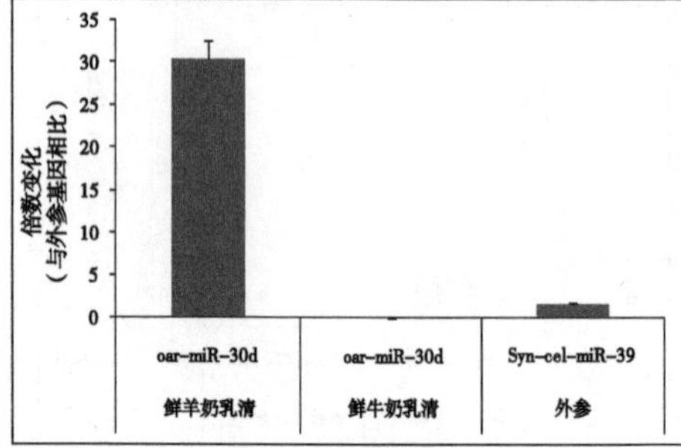

图 7-15　鲜羊奶乳清中 microRNA-30-5p、microRNA-200c、microRNA-21 和 microRNA-30d 相对表达量

Fig. 7-15　The relative expression of selected microRNA（miRNA）- microRNA-30-5p、microRNA-200c、microRNA-21 and microRNA-30d in fresh goat whey with normalization

注：采用果蝇 microNRA-39 进行标准化（To measure the levels of expression of miRNA in equal volumes of whey fraction from fresh goat whey，the quantitative PCR data obtained were normalized to spiked-in synthetic cel-miR-39）

异性，而 microRNA-200c 则在鲜牛奶乳清中检出。

由图 7-16 的结果可以看出，通过高通量测序所筛选出的表达丰度较高且在鲜羊奶乳清和鲜牛奶乳清中共有的 microRNA（microRNA-26a、microRNA-148a、microRNA-22-3p 和 microRNA-200b）均在鲜牛奶乳清和鲜羊奶乳清中检出，且在 microRNA-39 为外参的基础上进行校正，表达丰度均较高，其中的 microRNA-26a 和 microRNA-200b 在鲜牛奶乳清中的相对表达量高于鲜羊奶乳清，而

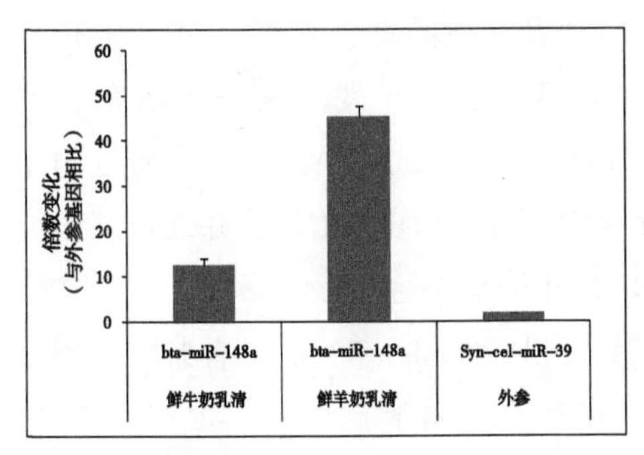

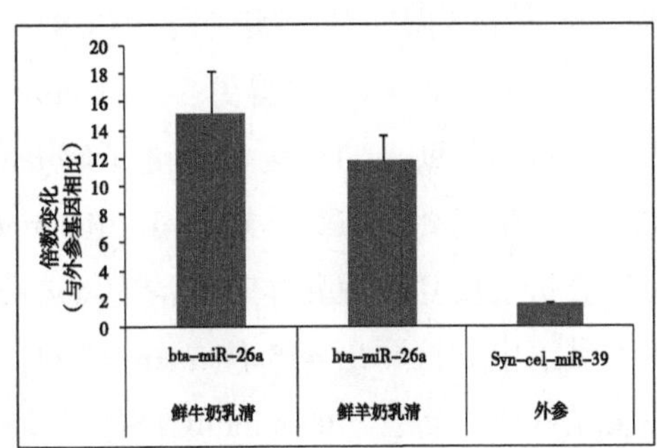

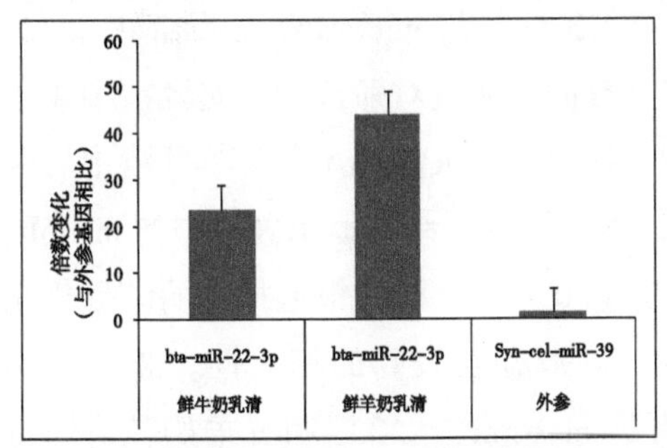

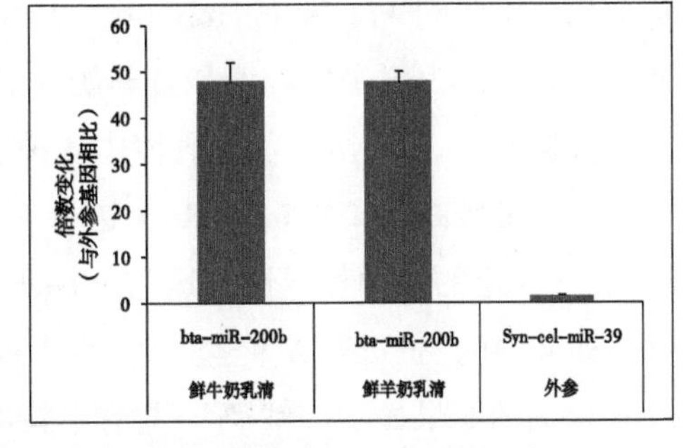

图 7-16　鲜牛羊奶乳清中 microRNA-26a、microRNA-148a、microRNA-22-3p 和 microRNA-200b 相对表达量

Fig. 7-16　The relative expression of selected microRNA（miRNA）-microRNA-26a、microRNA-148a、microRNA-22-3p and microRNA-200b in fresh cow and goat whey with normalization.

注：采用果蝇 microNRA-39 进行标准化（To measure the levels of expression of miRNA in equal volumes of whey fraction from fresh cow and goat whey, the quantitative PCR data obtained were normalized to spiked-in synthetic cel-miR-39）

microRNA-148a和microRNA-22-3p在鲜羊奶乳清中的相对表达量高于鲜牛奶乳清。

7.4 讨论

近几年，用于microRNA的检测的方法很多，如Northern Blot[187]、流式细胞术[188]、微阵列[189-190]，实时定量PCR[191-193]以及高通量测序[194-195]。基于高通量测序测定microRNA，使得更多的microRNA被发现，也扩展了microRNA的研究方向，进而帮助人们更为深入地了解microRNA。在本试验中，采用高通量测序检测了鲜牛奶乳清与鲜羊奶乳清中的microRNA及差异microRNA，并通过高通量测序分析了反复冻融乳样对乳清中microRNA种类及表达丰度的影响，其中在鲜牛奶乳清中共检测出415种microRNAs，其中381种microRNAs具有典型的颈环发夹结构匹配到已知的microRNAs，此外，还鉴定出34种新的microRNAs；在生鲜羊乳中检测到111种microRNAs，其中有13种新的microRNAs；而其中的29种microRNAs为鲜牛奶乳清和鲜羊奶乳清共有的。通过对筛选的牛奶乳清中4种表达量高的microRNA、羊奶乳清中4种表达量高的microRNA和牛羊乳清中均含有且表达量较高的4种microRNA进行RT-PCR验证，结果显示乳清中的microRNA存在一定的物种特异性，这与Weber等（2010）[152]研究报道指出的结果相一致，即乳中的microRNA种类和表达丰度与乳源存在较高的相关性。通过RT-PCR检测的结果显示，鲜奶牛乳清中特有的bta-miRNA200c和鲜羊奶乳清中特有的oar-miRNA200c分别在鲜羊奶乳清和鲜牛奶乳清中检测到，可能是由于这两个microRNA的碱基序列相近度较高，仅相差一个碱基（bta-microRNA200c：UAAUACUGCCGGGUAAUGAUGGA；oar-microRNA200c：UAAUACUGCCGGGUAAUGAUGG），可能在反转录和PCR扩增的时候出现错配，具体原因有待于进一步研究，并由此可以简单的确定bta-microRNA200c和oar-microRNA200c不可以作为鉴定鲜牛奶乳清和鲜羊奶乳清的特征性microRNA，而其他的microRNA均可以作为其各自特征性microRNA，用于后期的研究。

截至目前，有部分研究表明人乳[152-154]、牛乳[151,155-156]和猪乳[157-158]中均存

在 microRNA，且指出人乳中含有与免疫相关的 microRNA，且即使在酸性条件下也很稳定，可以直接传递给下一代，由肠道吸收并发挥其免疫功能[153-154]，而这些 microRNA 主要存在于乳中的外泌体中[153]。Izumi 等（2012）使用基因芯片和定量 PCR 研究牛乳的 miRNA，验证初乳和常乳之间的差异，初乳乳清中 RNA 的浓度高于常乳乳清 RNA 浓度。牛奶中共检测出 102 种 microRNA，与免疫相关的 microRNA 包括 miR-15b、miR-27b、miR-106b、miR-155 和 miR-223，与生长相关的 microRNA 包括 miR-27b、miR-34a 和 miR-130a，且初乳中这些 microRNA 浓度显著高于常乳[151]。Zhou 等（2012）分析了从出生至出生后 28 天泌乳猪乳中外泌体内的 microRNA，结果显示，其内富含与免疫相关的 microRNA，且初乳中的这些 microRNA 数量要显著高于常乳[157]。但针对羊乳中 microRNA 的检测的报道较少，牛乳中的 microRNA 的研究主要集中于不同泌乳阶段乳中 microRNA 的比较研究。本试验将新鲜的牛乳和羊乳进过多次离心制得乳清，进而采用高通量测序测定其各自所含的 microRNA，结果显示，牛乳和羊奶乳清中的 microRNA 的种类和表达丰度存在显著差异，表明不同物种乳中 microRNA 存在物种特异性，为进一步区分牛乳和羊乳提供一定的依据。

相对于鲜乳清，反复冻融乳清中的部分 microRNA 表达量上调或下调，揭示前者可能来源于冻融时乳中体细胞的损伤，使体细胞中的 microRNA 进入乳清中，后者可能是对冻融较敏感的 microRNA，可以作为新鲜乳潜在的标志性 microRNA 进行后续研究。但 Zhou 等（2012）研究分析了乳中外泌体内含的 microRNA 指出，外源合成的 miRNA 降解的非常快，而内源的 miRNA 则在长时间的室温下或反复冻融下依旧很稳定。此外，相比合成的 miRNA 在核糖核酸酶或 100℃ 培育 10 min 完全降解，而内源的 miRNA 在上述环境中依旧有很高的含量[153]，与本试验的结果存在差异，具体原因有待于进一步深入研究。

7.5 小结

（1）生鲜牛奶和山羊奶中 microRNA 的种类和含量存在显著差异。总的来说，生鲜牛奶中检测到 415 种 microRNAs，其中 381 种 microRNAs 具有典型的颈环发夹结构与已知的 microRNAs 匹配，此外，还鉴定出 34 种新的 microRNAs；在生鲜羊奶中检测到 111 种 microRNAs，其中有 13 种新的 microRNAs；而其中的 29 种 microRNAs 为鲜牛奶乳清和鲜羊奶乳清共有，并通过 RT-PCR 进行了验证。

（2）与新鲜牛乳清样本相比，反复冻融后 miR-1246、miR-135a、miR-15b、miR-223、miR-30b-5p、miR-342、miR-365-3p、miR-378b、miR-500、miR-664b 和 miR-885 表达上调，而 let-7a-5p、let-7c、let-7e、miR-122、miR-193a-5p、miR-25、miR-3432、miR-423-5p 和 miR-760-3p 表达下调。

（3）与新鲜羊乳清样本相比，反复冻融后 miR-23a、miR-23b、miR-30d 和 novel-7 表达上调，而 let-7g、miR-19b 和 miR-26b 表达下调。

8 结论与展望

8.1 总体结论

（1）本试验成功利用 RP-HPLC 分离和量化了奶牛乳、山羊乳、水牛乳及牦牛乳蛋白组分；并建立了奶牛乳、山羊乳、水牛乳和牦牛乳的 RP-HPLC 特征图谱，为进一步区分不同来源乳，甚至掺假乳提供一定的理论基础。

（2）在奶牛乳、水牛乳、牦牛乳和娟姗牛乳脂肪中 OBCFA 组成中含量最高的均是 *iso*-C15:0 和 C15:0；而在山羊乳中则是 C15:0 和 *anteiso*-C17:0；在马乳和骆驼乳中则是 *iso*-C15:0 和 *anteiso*-C17:0，均呈现一定的种属特异性。

（3）奶牛乳、水牛乳、牦牛乳、娟姗牛乳、山羊乳、骆驼乳和马乳中的 4 种常量元素（Na、Mg、K、Ca）和 5 种微量元素（Mn、Co、Zn、Se、Fe）含量差异显著；马乳中的 9 种元素，其中包括 4 种常量元素（Na、Mg、K、Ca）和 5 种微量元素（Mn、Co、Zn、Se、Fe）含量均显著低于其他物种乳。

（4）通过 Label-Free 定量蛋白质组学技术分离鉴定奶牛乳、水牛乳、牦牛乳、娟姗牛乳、山羊乳和骆驼乳中的 MFGM 蛋白，并确定这些蛋白的主要生物学功能和相关的调节路径；Cluster 聚类分析的结果显示，不同物种乳均具有一定的种属特异性。

（5）生鲜牛乳和山羊乳中 microRNA 的种类和含量存在显著差异。生鲜牛乳中检测到 415 种 microRNAs，其中 381 种 microRNAs 为已知的 microRNAs，其余的 34 种为新的 microRNAs；在生鲜羊乳中检测到 111 种 microRNAs，其中有 13 种新的 microRNAs；鲜牛奶乳清和鲜羊奶乳清共有的 microRNAs 为 29 种。

（6）鲜牛奶和鲜羊奶乳清与反复冻融牛乳和羊奶乳清中部分 mircoRNA 的表

达存在不同程度的上调或下调，表达下调的 microRNA 可能对冻融较敏感，可以作为新鲜乳潜在的标志性 microRNA 进行后续研究。

8.2 创新点及展望

8.2.1 创新点

本书中利用质谱技术结合蛋白质组学技术，构建了奶牛乳、水牛乳、牦牛乳、娟姗牛乳、山羊乳、骆驼乳和马乳中乳蛋白及 OBCFA 的特征性图谱，确定了不同畜种之间乳蛋白及 OBCFA 的特征性成分及其含量差异，并首次通过转录组学方法分析获得了新鲜奶牛与新鲜山羊乳清中特征性 microRNA 及反复冻融乳清与新鲜乳清样 microRNA 的表达差异性，发现了种属特异性以及对奶样冻融敏感的 microRNAs，为不同来源乳及掺假乳的区分提供了理论与方法学依据。

8.2.2 展望

本书较为系统地分析了不同畜种乳成分，并找出其成分的特征性，但今后还应在以下两方面继续开展研究：

（1）相等日粮营养条件下，单个畜种乳成分的特征性指标的探索；

（2）乳粉等商品化乳制品中 microRNA 的变化规律。

缩略语表

APC（antigen present cell）	抗原提呈细胞
FAO（Food and Agriculture Organization）	联合国粮食及农业组织
αs_1-CN（αs_1-casein）	αs_1-酪蛋白
αs_2-CN（αs_2-casein）	αs_2-酪蛋白
β-CN（β- casein）	β-酪蛋白
κ-CN（κ- casein）	κ-酪蛋白
β-LG（β- globulin）	β-乳球蛋白
α-LA（α-lactalbumin）	α-乳白蛋白
2-DE（2-dimensional electrophoresis）	二维凝胶电泳
MALDI-TOF-MS（Matrix-Assisted Laser Desorption/ Ionization Time of Flight Mass Spectrometry）	基质辅助激光解吸电离飞行时间质谱
SDS-PAGE（sodium dodecyl sulfate polyacrylamide gel electrophoresis）	十二烷基硫酸钠聚丙烯酰胺凝胶电泳
TLR（toll like receptor）	Toll 样受体
HPLC（high-performance liquid chromatography）	高效液相色谱
iTRAQ（isobaric tag for relative and absolute quantification）	同位素标记相对和绝对定量
RP-HPLC（reversed-phase high-performance liquid chromatography）	反相高效液相色谱
SDS-PAGE（sodium dodecyl sulfate polyacrylamide gel electrophoresis）	十二烷基硫酸钠聚丙烯酰胺凝胶电泳
TLC（thin-Layer Chromatography）	薄层色谱
OBCFA（odd and branched chain fatty acids）	奇数碳链支链脂肪酸
BCFA（branched chain fatty acids）	支链脂肪酸

（续表）

FA（fatty acid）	脂肪酸
ICP-MS（Inductively Coupled Plasma Mass Spectrometry）	电感耦合等离子体质谱
GC-MS（Gas Chromatography-Mass Spectrometer）	气相色谱-质谱联用仪
GdnHCl（Guanidine hydrochloride）	盐酸胍
PCA（principal component analysis）	主成分分析
UHT（ultra-high-temperature）	超高温灭菌
LC-MS（liquid chromatography-mass spectrometry）	液相色谱或串联质谱
GC（gas chromatograph）	气相色谱

参考文献

[1] Mcguire E. Ruth goes home: An adult's use of human milk [J]. Breastfeeding Review, 2012, 20 (3): 44.

[2] Gao Z, Yin J, Zhang J, et al. Butyrate improves insulin sensitivity and increases energy expenditure in mice [J]. Diabetes, 2009, 58 (7): 1 509-1 517.

[3] Tantibhedhyangkul P, Hashim S A. Medium-chain triglyceride feeding in premature infants: effects on fat and nitrogen absorption [J]. Pediatrics, 1975, 55 (3): 359-370.

[4] Haug A, Hostmark A T, Harstad O M. Bovine milk in human nutrition-a review [J]. Lipids Health Dis, 2007, 6 (1): 25.

[5] Lock A L, Bauman D E. Modifying milk fat composition of dairy cows to enhance fatty acids beneficial to human health [J]. Lipids, 2004, 39 (12): 1 197-1 206.

[6] Medhammar E, Wijesinha Bettoni R, Stadlmayr B, et al. Composition of milk from minor dairy animals and buffalo breeds: a biodiversity perspective [J]. Journal of the Science of Food and Agriculture, 2012, 92 (3): 445-474.

[7] Tesse R, Paglialunga C, Braccio S, et al. Adequacy and tolerance to ass's milk in an Italian cohort of children with cow's milk allergy [J]. Italian journal of pediatrics, 2009, 35 (19).

[8] Silk T M, Guo M, Haenlein G F, et al. Yak milk [J]. Handbook of Milk of Non-Bovine Mammals, 2006: 345-353.

[9] Gaucheron F. The minerals of milk [J]. Reproduction Nutrition Development, 2005, 45 (4): 473-484.

[10] Guéguen L, Pointillart A. The bioavailability of dietary calcium [J]. Journal of the American College of Nutrition, 2000, 19 (sup2): 119S-136S.

[11] Barłowska J, Szwajkowska M, Litwińczuk Z, et al. Nutritional value and technological suitability of milk from various animal species used for dairy production [J]. Comprehensive Reviews in Food Science and Food Safety, 2011, 10 (6): 291-302.

[12] Park Y W. Rheological characteristics of goat and sheep milk [J]. Small Ruminant Research, 2007, 68 (1): 73-87.

[13] Raynal-Ljutovac K, Lagriffoul G, Paccard P, et al. Composition of goat and sheep milk products: An update [J]. Small ruminant research, 2008, 79 (1): 57-72.

[14] Smiddy M A, Huppertz T, van Ruth S M. Triacylglycerol and melting profiles of milk fat from several species [J]. International Dairy Journal, 2012, 24 (2): 64-69.

[15] Blasi F, Montesano D, De Angelis M, et al. Results of stereospecific analysis of triacylglycerol fraction from donkey, cow, ewe, goat and buffalo milk [J]. Journal of food composition and analysis, 2008, 21 (1): 1-7.

[16] Hinz K, O'Connor P M, Huppertz T, et al. Comparison of the principal proteins in bovine, caprine, buffalo, equine and camel milk [J]. Journal of Dairy Research, 2012, 79 (02): 185-191.

[17] Barłowska J, Król J. Milk protein polymorphism as markers of production traits in dairy and meat cattle. [J]. Medycyna Weterynaryjna, 2006, 62 (1): 6-10.

[18] D'Auria E, Agostoni C, Giovannini M, et al. Proteomic evaluation of

milk from different mammalian species as a substitute for breast milk [J]. Acta paediatrica, 2005, 94 (12): 1 708-1 713.

[19] Aslam M, Jimenez-Flores R, Kim H Y, et al. Two-dimensional electrophoretic analysis of proteins of bovine mammary gland secretions collected during the dry period [J]. Journal of dairy science, 1994, 77 (6): 1 529-1 536.

[20] Smolenski G, Haines S, Kwan F Y, et al. Characterisation of host defence proteins in milk using a proteomic approach [J]. Journal of proteome research, 2007, 6 (1): 207-215.

[21] Yamada M, Murakami K, Wallingford J C, et al. Identification of low abundance proteins of bovine colostral and mature milk using two dimensional electrophoresis followed by microsequencing and mass spectrometry [J]. Electrophoresis, 2002, 23 (7-8): 1 153-1 160.

[22] Holland J W, Deeth H C, Alewood P F. Proteomic analysis of κ-casein micro-heterogeneity [J]. Proteomics, 2004, 4 (3): 743-752.

[23] Reinhardt T A, Lippolis J D. Developmental changes in the milk fat globule membrane proteome during the transition from colostrum to milk [J]. Journal of dairy science, 2008, 91 (6): 2 307-2 318.

[24] Vanderghem C, Blecker C, Danthine S, et al. Proteome analysis of the bovine milk fat globule: Enhancement of membrane purification [J]. International dairy journal, 2008, 18 (9): 885-893.

[25] Affolter M, Grass L, Vanrobaeys F, et al. Qualitative and quantitative profiling of the bovine milk fat globule membrane proteome [J]. Journal of proteomics, 2010, 73 (6): 1 079-1 088.

[26] Bramanti E, Sortino C, Raspi G. New chromatographic method for separation and determination of denatured αs_1-, αs_2-, β-and κ-caseins by hydrophobic interaction chromatography [J]. Journal of Chromatography A, 2002, 958 (1): 157-166.

[27] Bonizzi I, Buffoni J N, Feligini M. Quantification of bovine casein fractions by direct chromatographic analysis of milk. Approaching the application to a real production context [J]. Journal of Chromatography A, 2009, 1216 (1): 165–168.

[28] Chen R K, Chang L W, Chung Y Y, et al. Quantification of cow milk adulteration in goat milk using high–performance liquid chromatography with electrospray ionization mass spectrometry [J]. Rapid communications in mass spectrometry, 2004, 18 (10): 1 167–1 171.

[29] Feligini M, Bonizzi I, Buffoni J N, et al. Identification and quantification of αS_1, αS_2, β, and κ–caseins in water buffalo milk by reverse phase–high performance liquid chromatography and mass spectrometry [J]. Journal of agricultural and food chemistry, 2009, 57 (7): 2 988–2 992.

[30] Buffoni J N, Bonizzi I, Pauciullo A, et al. Characterization of the major whey proteins from milk of Mediterranean water buffalo (Bubalus bubalis) [J]. Food Chemistry, 2011, 127 (4): 1 515–1 520.

[31] Bonfatti V, Grigoletto L, Cecchinato A, et al. Validation of a new reversed–phase high–performance liquid chromatography method for separation and quantification of bovine milk protein genetic variants [J]. Journal of Chromatography A, 2008, 1195 (1): 101–106.

[32] Bonfatti V, Giantin M, Rostellato R, et al. Separation and quantification of water buffalo milk protein fractions and genetic variants by RP–HPLC [J]. Food chemistry, 2013, 136 (2): 364–367.

[33] Martino F A R, Sánchez M L F, Sanz–Medel A. The potential of double focusing–ICP–MS for studying elemental distribution patterns in whole milk, skimmed milk and milk whey of different milks [J]. Analytica chimica acta, 2001, 442 (2): 191–200.

[34] Bonfatti V, Di Martino G, Cecchinato A, et al. Effects of β–κ–casein (CSN2–CSN3) haplotypes, β–lactoglobulin (BLG) genotypes, and

detailed protein composition on coagulation properties of individual milk of Simmental cows [J]. Journal of dairy science, 2010, 93 (8): 3 809–3 817.

[35] Farrell Jr H M, Jimenez–Flores R, Bleck G T, et al. Nomenclature of the proteins of cows' milk—sixth revision [J]. Journal of dairy science, 2004, 87 (6): 1 641–1 674.

[36] Moatsou G, Vamvakaki A, Mollé D, et al. Protein composition and polymorphism in the milk of Skopelos goats [J]. Le Lait, 2006, 86 (5): 345–357.

[37] Ochirkhuyag B, Chobert J M, Dalgalarrondo M, et al. Characterization of whey proteins from Mongolian yak, khainak, and bactrian camel [J]. Journal of Food biochemistry, 1998, 22 (2): 105–124.

[38] Li H, Ma Y, Dong A, et al. Protein composition of yak milk [J]. Dairy Science & Technology, 2010, 90 (1): 111–117.

[39] Zicarelli L. Buffalo milk: its properties, dairy yield and mozzarella production [J]. Veterinary research communications, 2004, 28 (Supplement 1): 127–135.

[40] Inayat S, Arain M A, Khaskheli M, et al. Study on the production and quality improvement of soft unripened cheese made from buffalo milk as compared with camel milk [J]. Italian Journal of Animal Science, 2010, 6 (2s): 1 115–1 119.

[41] Leitner G, Merin U, Silanikove N. Changes in milk composition as affected by subclinical mastitis in goats [J]. Journal of Dairy Science, 2004, 87 (6): 1 719–1 726.

[42] Guo H Y, Pang K, Zhang X Y, et al. Composition, physiochemical properties, nitrogen fraction distribution, and amino acid profile of donkey milk [J]. Journal of dairy science, 2007, 90 (4): 1 635–1 643.

[43] Amigo L, Recio I, Ramos M. Genetic polymorphism of ovine milk

proteins: its influence on technological properties of milk-a review [J]. International Dairy Journal, 2000, 10 (3): 135-149.

[44] Bobe G, Lindberg G L, Freeman A E, et al. Short Communication: Composition of Milk Protein and Milk Fatty Acids Is Stable for Cows Differing in Genetic Merit for Milk Production [J]. Journal of dairy science, 2007, 90 (8): 3 955-3 960.

[45] Visser S, Slangen C J, Rollema H S. Phenotyping of bovine milk proteins by reversed-phase high-performance liquid chromatography [J]. Journal of Chromatography A, 1991, 548: 361-370.

[46] Addeo F, Nicolai M A, Chianese L, et al. A control method to detect bovine milk in ewe and water buffalo cheese using immunoblotting [J]. Milchwissenschaft, 1995, 50 (2): 83-85.

[47] Recio I, Amigo L, López-Fandiño R. Assessment of the quality of dairy products by capillary electrophoresis of milk proteins [J]. Journal of Chromatography B: Biomedical Sciences and Applications, 1997, 697 (1): 231-242.

[48] Cartoni G P, Coccioli F, Jasionowska R, et al. Determination of cow milk in ewe milk and cheese by capillary electrophoresis of the whey protein fractions [J]. Italian journal of food science, 1998, 10 (4): 317-327.

[49] De Block J, Merchiers M, Van Renterghem R. Capillary electrophoresis of the whey protein fraction of milk powders. A possible method for monitoring storage conditions [J]. International dairy journal, 1998, 8 (9): 787-792.

[50] Bravo F I, Molina E, López-Fandiño R. Effect of the high-pressure-release phase on the protein composition of the soluble milk fraction [J]. Journal of dairy science, 2012, 95 (11): 6 293-6 299.

[51] Garbis S, Lubec G, Fountoulakis M. Limitations of current proteomics

technologies [J]. Journal of Chromatography A, 2005, 1077 (1): 1-18.

[52] O Donnell R, Holland J W, Deeth H C, et al. Milk proteomics [J]. International Dairy Journal, 2004, 14 (12): 1 013-1 023.

[53] Yang Y, Bu D, Zhao X, et al. Proteomic Analysis of Cow, Yak, Buffalo, Goat and Camel Milk Whey Proteins: Quantitative Differential Expression Patterns [J]. Journal of proteome research, 2013, 12 (4): 1 660-1 667.

[54] Bordin G, Cordeiro Raposo F, De la Calle B, et al. Identification and quantification of major bovine milk proteins by liquid chromatography [J]. Journal of chromatography A, 2001, 928 (1): 63-76.

[55] Careri M, Mangia A. Analysis of food proteins and peptides by chromatography and mass spectrometry [J]. Journal of Chromatography A, 2003, 1000 (1): 609-635.

[56] Aschaffenburg R, Sen A. Comparison of the caseins of buffalo's and cow's milk [J]. 1963.

[57] Trujillo A J, Casals I, Guamis B. Analysis of major caprine milk proteins by reverse - phase high - performance liquid chromatography and electrospray ionization-mass spectrometry [J]. Journal of dairy science, 2000, 83 (1): 11-19.

[58] Cozzolino R, Passalacqua S, Salemi S, et al. Identification of adulteration in water buffalo mozzarella and in ewe cheese by using whey proteins as biomarkers and matrix-assisted laser desorption/ionization mass spectrometry [J]. Journal of mass spectrometry, 2002, 37 (9): 985-991.

[59] Somma A, Ferranti P, Addeo F, et al. Peptidomic approach based on combined capillary isoelectric focusing and mass spectrometry for the characterization of the plasmin primary products from bovine and water buffalo β- casein [J]. Journal of Chromatography A, 2008, 1192 (2):

294-300.

[60] Mcmanaman J L, Neville M C. Mammary physiology and milk secretion [J]. Advanced drug delivery reviews, 2003, 55 (5): 629-641.

[61] Fong B Y, Norris C S. Quantification of milk fat globule membrane proteins using selected reaction monitoring mass spectrometry [J]. Journal of agricultural and food chemistry, 2009, 57 (14): 6 021-6 028.

[62] Cavaletto M, Giuffrida M G, Conti A. Milk fat globule membrane components—a proteomic approach [M]. Bioactive Components of Milk, Springer, 2008: 129-141.

[63] Dewettinck K, Rombaut R, Thienpont N, et al. Nutritional and technological aspects of milk fat globule membrane material [J]. International dairy journal, 2008, 18 (5): 436-457.

[64] Spitsberg V L. Invited Review: Bovine Milk Fat Globule Membrane as a Potential Nutraceutical [J]. Journal of dairy science, 2005, 88 (7): 2 289-2 294.

[65] Mañá P, Goodyear M, Bernard C, et al. Tolerance induction by molecular mimicry: prevention and suppression of experimental autoimmune encephalomyelitis with the milk protein butyrophilin [J]. International immunology, 2004, 16 (3): 489-499.

[66] Martin H M, Hancock J T, Salisbury V, et al. Role of xanthine oxidoreductase as an antimicrobial agent [J]. Infection and immunity, 2004, 72 (9): 4 933-4 939.

[67] Hancock J T, Salisbury V, Ovejero-Boglione M C, et al. Antimicrobial properties of milk: dependence on presence of xanthine oxidase and nitrite [J]. Antimicrobial agents and chemotherapy, 2002, 46 (10): 3 308-3 310.

[68] Charlwood J, Hanrahan S, Tyldesley R, et al. Use of proteomic method-

ology for the characterization of human milk fat globular membrane proteins [J]. Analytical biochemistry, 2002, 301 (2): 314-324.

[69] Liao Y, Alvarado R, Phinney B, et al. Proteomic characterization of human milk fat globule membrane proteins during a 12 month lactation period [J]. Journal of proteome research, 2011, 10 (8): 3 530-3 541.

[70] Cavaletto M, Giuffrida M G, Conti A. The proteomic approach to analysis of human milk fat globule membrane [J]. Clinica chimica acta, 2004, 347 (1): 41-48.

[71] Reinhardt T A, Lippolis J D. Bovine milk fat globule membrane proteome [J]. Journal of Dairy Research, 2006, 73 (4): 406-416.

[72] Bianchi L, Puglia M, Landi C, et al. Solubilization methods and reference 2-DE map of cow milk fat globules [J]. Journal of proteomics, 2009, 72 (5): 853-864.

[73] Dickow J A, Larsen L B, Hammershøj M, et al. Cooling causes changes in the distribution of lipoprotein lipase and milk fat globule membrane proteins between the skim milk and cream phase [J]. Journal of dairy science, 2011, 94 (2): 646-656.

[74] Reinhardt T A, Sacco R E, Nonnecke B J, et al. Bovine milk proteome: Quantitative changes in normal milk exosomes, milk fat globule membranes and whey proteomes resulting from Staphylococcus aureus mastitis [J]. Journal of proteomics, 2013, 82: 141-154.

[75] Lu J, Boeren S, de Vries S C, et al. Filter-aided sample preparation with dimethyl labeling to identify and quantify milk fat globule membrane proteins [J]. Journal of proteomics, 2011, 75 (1): 34-43.

[76] Nikkhah A. Equidae, camel, and yak milks as functional foods: a review [J]. Journal of Nutrition & Food Sciences, 2011.

[77] Cebo C, Caillat H, Bouvier F, et al. Major proteins of the goat milk fat globule membrane [J]. Journal of dairy science, 2010, 93 (3):

868-876.

[78] Zamora A, Guamis B, Trujillo A J. Protein composition of caprine milk fat globule membrane [J]. Small ruminant research, 2009, 82 (2): 122-129.

[79] D'Ambrosio C, Arena S, Salzano A M, et al. A proteomic characterization of water buffalo milk fractions describing PTM of major species and the identification of minor components involved in nutrient delivery and defense against pathogens [J]. Proteomics, 2008, 8 (17): 3 657-3 666.

[80] Pisanu S, Ghisaura S, Pagnozzi D, et al. The sheep milk fat globule membrane proteome [J]. Journal of proteomics, 2011, 74 (3): 350-358.

[81] Cebo C, Rebours E, Henry C, et al. Identification of major milk fat globule membrane proteins from pony mare milk highlights the molecular diversity of lactadherin across species [J]. Journal of dairy science, 2012, 95 (3): 1 085-1 098.

[82] Barello C, Garoffo L P, Montorfano G, et al. Analysis of major proteins and fat fractions associated with mare's milk fat globules [J]. Molecular nutrition & food research, 2008, 52 (12): 1 448-1 456.

[83] Saadaoui B, Henry C, Khorchani T, et al. Proteomics of the milk fat globule membrane from Camelus dromedarius [J]. Proteomics, 2013, 13 (7): 1 180-1 184.

[84] Murgiano L, Timperio A M, Zolla L, et al. Comparison of milk fat globule membrane (MFGM) proteins of chianina and holstein cattle breed milk samples through proteomics methods [J]. Nutrients, 2009, 1 (2): 302-315.

[85] Murgiano L, D'Alessandro A, Zolla L, et al. Comparison of Milk Fat Globule Membrane (MFGM) proteins in milk samples of Chianina and Holstein cattle breeds across three lactation phases through 2D IEF SDS PAGE—A preliminary study [J]. Food Research International, 2013,

54 (1): 1 280-1 286.

[86] Cebo C, Martin P. Inter-species comparison of milk fat globule membrane proteins highlights the molecular diversity of lactadherin [J]. International Dairy Journal, 2012, 24 (2): 70-77.

[87] Spertino S, Cipriani V, De Angelis C, et al. Proteome profile and biological activity of caprine, bovine and human milk fat globules [J]. Molecular BioSystems, 2012, 8 (4): 967-974.

[88] El-Zeini H M. Microstructure, rheological and geometrical properties of fat globules of milk from different animal species [J]. Polish journal of food and nutrition sciences, 2006, 15 (2): 147.

[89] Attaie R, Richter R L. Size distribution of fat globules in goat milk [J]. Journal of Dairy Science, 2000, 83 (5): 940-944.

[90] Jensen R G. The composition of bovine milk lipids: January 1995 to December 2000 [J]. Journal of Dairy Science, 2002, 85 (2): 295-350.

[91] Iverson J L, Sheppard A J. Detection of adulteration in cow, goat, and sheep cheeses utilizing gas-liquid chromatographic fatty acid data [J]. Journal of Dairy Science, 1989, 72 (7): 1 707-1 712.

[92] Haddad I, Mozzon M, Strabbioli R, et al. Stereospecific analysis of triacylglycerols in camel (Camelus dromedarius) milk fat [J]. International dairy journal, 2010, 20 (12): 863-867.

[93] Parodi P W. Dairy product consumption and the risk of breast cancer [J]. Journal of the American College of Nutrition, 2005, 24 (sup6): 556S-568S.

[94] Vlaeminck B, Fievez V, Cabrita A, et al. Factors affecting odd-and branched-chain fatty acids in milk: a review [J]. Animal Feed Science and Technology, 2006, 131 (3): 389-417.

[95] Schreiberová O, Krulikovská T, Sigler K, et al. RP-HPLC/MS-APCI

analysis of branched chain TAG prepared by precursor - directed biosynthesis with Rhodococcus erythropolis [J]. Lipids, 2010, 45 (8): 743-756.

[96] Kuhajda F P. Fatty-acid synthase and human cancer: new perspectives on its role in tumor biology [J]. Nutrition, 2000, 16 (3): 202-208.

[97] Wongtangtintharn S, Oku H, Iwasaki H, et al. Effect of branched-chain fatty acids on fatty acid biosynthesis of human breast cancer cells. [J]. Journal of nutritional science and vitaminology, 2004, 50 (2): 137-143.

[98] Enser M. chemistry, biochemistry and nutritional importance of animal fats [J]. Proceedings-Easter School in Agricultural Science, University of Nottingham, 1984.

[99] Diedrich M, Henschel K P. The natural occurrence of unusual fatty acids. Part 1. Odd numbered fatty acids [J]. Food/Nahrung, 1990, 34 (10): 935-943.

[100] Polidori P, Maggi G L, Moretti V M, et al. A note on the effect of use of bovine somatotropin on the fatty acid composition of the milk fat in dairy cows [J]. ANIMAL PRODUCTION - GLASGOW -, 1993, 57: 319.

[101] Jenkins T C. Butylsoyamide protects soybean oil from ruminal biohydrogenation: effects of butylsoyamide on plasma fatty acids and nutrient digestion in sheep. [J]. Journal of animal science, 1995, 73 (3): 818-823.

[102] Rojas A, Lopez-Bote C, Rota A, et al. Fatty acid composition of Verata goat kids fed either goat milk or commercial milk replacer [J]. Small Ruminant Research, 1994, 14 (1): 61-66.

[103] Käkelä R, Hyvärinen H, Vainiotalo P. Unusual fatty acids in the depot fat of the Canadian beaver (Castor canadensis) [J]. Comparative Bio-

chemistry and Physiology Part B: Biochemistry and Molecular Biology, 1996, 113 (3): 625-629.

[104] Keeney M, Katz I, Allison M J. On the probable origin of some milk fat acids in rumen microbial lipids [J]. Journal of the American Oil Chemists Society, 1962, 39 (4): 198-201.

[105] Craninx M, Steen A, Van Laar H, et al. Effect of lactation stage on the odd-and branched-chain milk fatty acids of dairy cattle under grazing and indoor conditions [J]. Journal of dairy science, 2008, 91 (7): 2 662-2 677.

[106] Dewhurst R J, Moorby J M, Vlaeminck B, et al. Apparent recovery of duodenal odd-and branched-chain fatty acids in milk of dairy cows [J]. Journal of dairy science, 2007, 90 (4): 1 775-1 780.

[107] Fievez V, Colman E, Castro - Montoya J M, et al. Milk odd - and branched-chain fatty acids as biomarkers of rumen function—An update [J]. Animal Feed Science and Technology, 2012, 172 (1): 51-65.

[108] Alonso L, Fontecha J, Lozada L, et al. Fatty Acid Composition of Caprine Milk: Major, Branched - Chain, and Trans Fatty Acids [J]. Journal of dairy science, 1999, 82 (5): 878-884.

[109] Massart-Leen A M, De Pooter H, Decloedt M, et al. Composition and variability of the branched-chain fatty acid fraction in the milk of goats and cows [J]. Lipids, 1981, 16 (5): 286-292.

[110] Devle H, Vetti I, Naess Andresen C F, et al. A comparative study of fatty acid profiles in ruminant and non-ruminant milk [J]. European Journal of Lipid Science and Technology, 2012, 114 (9): 1 036-1 043.

[111] Liu H N, Ren F Z, Jiang L, et al. Short communication: Fatty acid profile of yak milk from the Qinghai-Tibetan Plateau in different seasons and for different parities [J]. Journal of dairy science, 2011, 94 (4):

1 724-1 731.

[112] Ding L, Wang Y, Kreuzer M, et al. Seasonal variations in the fatty acid profile of milk from yaks grazing on the Qinghai-Tibetan plateau [J]. The Journal of dairy research, 2013, 80 (4): 410.

[113] Neupaney D, Ishioroshi M, Samejima K. Study on some functional and compositional properties of yak butter lipid [J]. Animal Science Journal, 2003, 74 (5): 391-397.

[114] White S L, Bertrand J A, Wade M R, et al. Comparison of fatty acid content of milk from Jersey and Holstein cows consuming pasture or a total mixed ration [J]. Journal of Dairy Science, 2001, 84 (10): 2 295-2 301.

[115] Auldist M J, Johnston K A, White N J, et al. A comparison of the composition, coagulation characteristics and cheesemaking capacity of milk from Friesian and Jersey dairy cows [J]. Journal of Dairy Research, 2004, 71 (01): 51-57.

[116] Curadi M C, Leotta R, Contarini G, et al. Milk fatty acids from different horse breeds compared with cow, goat and human milk. [J]. Macedonian Journal of Animal Science, 2012, 2 (1): 79-82.

[117] Karray N, Lopez C, Ollivon M, et al. La matière grasse du lait de dromadaire: composition, microstructure et polymorphisme. Une revue [J]. Oléagineux, Corps Gras, Lipides, 2005, 12 (5): 439-446.

[118] Konuspayeva G, Lemarie É, Faye B, et al. Fatty acid and cholesterol composition of camel's (Camelus bactrianus, Camelus dromedarius and hybrids) milk in Kazakhstan [J]. Dairy science and technology, 2008, 88 (3): 327-340.

[119] Herwig N, Stephan K, Panne U, et al. Multi-element screening in milk and feed by SF-ICP-MS [J]. Food chemistry, 2011, 124 (3): 1 223-1 230.

[120] Sola-Larrañaga C, Navarro-Blasco I. Chemometric analysis of minerals and trace elements in raw cow milk from the community of Navarra, Spain [J]. Food Chemistry, 2009, 112 (1): 189-196.

[121] Toledo Á, Burlingame B. Biodiversity and nutrition: A common path toward global food security and sustainable development [J]. Journal of Food Composition and Analysis, 2006, 19 (6): 477-483.

[122] Burlingame B, Charrondiere R, Mouille B. Food composition is fundamental to the cross-cutting initiative on biodiversity for food and nutrition [J]. Journal of food composition and analysis, 2009, 22 (5): 361-365.

[123] Ataro A, Mccrindle R I, Botha B M, et al. Quantification of trace elements in raw cow's milk by inductively coupled plasma mass spectrometry (ICP-MS) [J]. Food chemistry, 2008, 111 (1): 243-248.

[124] Ayar A, Sert D, Akın N. The trace metal levels in milk and dairy products consumed in middle Anatolia—Turkey [J]. Environmental monitoring and assessment, 2009, 152 (1-4): 1-12.

[125] Pilarczyk R, Wójcik J, Czerniak P, et al. Concentrations of toxic heavy metals and trace elements in raw milk of Simmental and Holstein-Friesian cows from organic farm [J]. Environmental monitoring and assessment, 2013, 185 (10): 8 383-8 392.

[126] Dobrzański Z, Kołacz R, Górecka H, et al. The Content of Microelements and Trace Elements in Raw Milk from Cows in the Silesian Region [J]. Polish Journal of Environmental Studies, 2005, 14 (5) .

[127] D Ilio S, Petrucci F, D Amato M, et al. Method validation for determination of arsenic, cadmium, chromium and lead in milk by means of dynamic reaction cell inductively coupled plasma mass spectrometry [J]. Analytica chimica acta, 2008, 624 (1): 59-67.

[128] Khan Z I, Ashraf M, Hussain A, et al. Concentrations of minerals in

milk of sheep and goats grazing similar pastures in a semiarid region of Pakistan [J]. Small ruminant research, 2006, 65 (3): 274–278.

[129] Güler Z. Levels of 24 minerals in local goat milk, its strained yoghurt and salted yoghurt (tuzlu yoğurt) [J]. Small ruminant research, 2007, 71 (1): 130–137.

[130] Krachler M, J Irgolic K. The potential of inductively coupled plasma mass spectrometry (ICP–MS) for the simultaneous determination of trace elements in whole blood, plasma and serum [J]. Journal of trace elements in medicine and biology, 1999, 13 (3): 157–169.

[131] Lyon T D, Fell G S, Hutton R C, et al. Evaluation of inductively coupled argon plasma mass spectrometry (ICP–MS) for simultaneous multi-element trace analysis in clinical chemistry [J]. J. Anal. At. Spectrom, 1988, 3 (1): 265–271.

[132] Muñiz C S, Gayón J M M, Alonso J I G, et al. Speciation of essential elements in human serum using anion–exchange chromatography coupled to post–column isotope dilution analysis with double focusing ICP–MS [J]. Journal of Analytical Atomic Spectrometry, 2001, 16 (6): 587–592.

[133] Llorent–Martínez E J, de Córdova M L, Ruiz–Medina A, et al. Analysis of 20 trace and minor elements in soy and dairy yogurts by ICP–MS [J]. Microchemical Journal, 2012, 102: 23–27.

[134] Lagos–Quintana M, Rauhut R, Lendeckel W, et al. Identification of novel genes coding for small expressed RNAs [J]. Science, 2001, 294 (5543): 853–858.

[135] Lau N C, Lim L P, Weinstein E G, et al. An abundant class of tiny RNAs with probable regulatory roles in Caenorhabditis elegans [J]. Science, 2001, 294 (5543): 858–862.

[136] Lee R C, Ambros V. An extensive class of small RNAs in Caenorhabditis

elegans [J]. Science, 2001, 294 (5543): 862-864.

[137] Yang W J, Yang D D, Na S, et al. Dicer is required for embryonic angiogenesis during mouse development [J]. Journal of Biological Chemistry, 2005, 280 (10): 9 330-9 335.

[138] van Rooij E. The art of microRNA research [J]. Circulation research, 2011, 108 (2): 219-234.

[139] Cai X, Hagedorn C H, Cullen B R. Human microRNAs are processed from capped, polyadenylated transcripts that can also function as mRNAs [J]. Rna, 2004, 10 (12): 1 957-1 966.

[140] Lee Y, Kim M, Han J, et al. MicroRNA genes are transcribed by RNA polymerase II [J]. The EMBO journal, 2004, 23 (20): 4 051-4 060.

[141] Denli A M, Tops B B, Plasterk R H, et al. Processing of primary microRNAs by the Microprocessor complex [J]. Nature, 2004, 432 (7014): 231-235.

[142] Gregory R I, Yan K, Amuthan G, et al. The Microprocessor complex mediates the genesis of microRNAs [J]. Nature, 2004, 432 (7014): 235-240.

[143] Lee Y, Ahn C, Han J, et al. The nuclear RNase III Drosha initiates microRNA processing [J]. nature, 2003, 425 (6956): 415-419.

[144] Yi R, Qin Y, Macara I G, et al. Exportin-5 mediates the nuclear export of pre-microRNAs and short hairpin RNAs [J]. Genes & development, 2003, 17 (24): 3 011-3 016.

[145] Chendrimada T P, Gregory R I, Kumaraswamy E, et al. TRBP recruits the Dicer complex to Ago2 for microRNA processing and gene silencing [J]. Nature, 2005, 436 (7051): 740-744.

[146] Khvorova A, Reynolds A, Jayasena S D. Functional siRNAs and miRNAs exhibit strand bias [J]. Cell, 2003, 115 (2): 209-216.

[147] Davis T A, Nguyen H V, Garcia - Bravo R, et al. Amino acid composition of the milk of some mammalian species changes with stage of lactation [J]. British journal of nutrition, 1994, 72 (06): 845-853.

[148] Jensen R G. Handbook of milk composition [M]. Academic Press, 1995.

[149] Kosaka N, Iguchi H, Ochiya T. Circulating microRNA in body fluid: a new potential biomarker for cancer diagnosis and prognosis [J]. Cancer science, 2010, 101 (10): 2 087-2 092.

[150] Lewis B P, Burge C B, Bartel D P. Conserved seed pairing, often flanked by adenosines, indicates that thousands of human genes are microRNA targets [J]. cell, 2005, 120 (1): 15-20.

[151] Izumi H, Kosaka N, Shimizu T, et al. Bovine milk contains microRNA and messenger RNA that are stable under degradative conditions [J]. Journal of dairy science, 2012, 95 (9): 4 831-4 841.

[152] Weber J A, Baxter D H, Zhang S, et al. The microRNA spectrum in 12 body fluids [J]. Clinical chemistry, 2010, 56 (11): 1 733-1 741.

[153] Zhou Q, Li M, Wang X, et al. Immune - related microRNAs are abundant in breast milk exosomes [J]. International journal of biological sciences, 2012, 8 (1): 118.

[154] Kosaka N, Izumi H, Sekine K, et al. microRNA as a new immune-regulatory agent in breast milk [J]. Silence, 2010, 1 (1): 1-8.

[155] Chen X, Gao C, Li H, et al. Identification and characterization of microRNAs in raw milk during different periods of lactation, commercial fluid, and powdered milk products [J]. Cell research, 2010, 20 (10): 1 128-1 137.

[156] Hata T, Murakami K, Nakatani H, et al. Isolation of bovine milk - derived microvesicles carrying mRNAs and microRNAs [J]. Biochemical and biophysical research communications, 2010, 396 (2): 528-533.

[157] Gu Y, Li M, Wang T, et al. Lactation-related microRNA expression

profiles of porcine breast milk exosomes [J]. PloS one, 2012, 7 (8): e43691.

[158] Gu Y R, Liang Y, Gong J J, et al. Suitable internal control microRNA genes for measuring miRNA abundance in pig milk during different lactation periods [J]. Genetics and Molecular Research, 2012, 11 (3): 2 506-2 512.

[159] Hallén E, Wedholm A, Andrén A, et al. Effect of β-casein, κ-casein and β-lactoglobulin genotypes on concentration of milk protein variants [J]. Journal of Animal Breeding and genetics, 2008, 125 (2): 119-129.

[160] Mercier J C, Addeo F, Pelissier J P. Primary structure of the casein macropeptide of caprine kappa casein [J]. Biochimie, 1975, 58 (11-12): 1 303-1 310.

[161] Ferreira I M, Caçote H. Detection and quantification of bovine, ovine and caprine milk percentages in protected denomination of origin cheeses by reversed-phase high-performance liquid chromatography of beta-lactoglobulins [J]. Journal of Chromatography A, 2003, 1015 (1): 111-118.

[162] Lopez-Fandi O R, Acedo M I, Ramos M. Comparative study by HPLC of caseinomacropeptides from cows', ewes' and goats' milk [J]. Journal of dairy research, 1993, 60: 117.

[163] Bramanti E, Sortino C, Onor M, et al. Separation and determination of denatured αs_1-, αs_2-, β-and κ-caseins by hydrophobic interaction chromatography in cows', ewes' and goats' milk, milk mixtures and cheeses [J]. Journal of chromatography A, 2003, 994 (1): 59-74.

[164] López Expósito I, Recio I. Antibacterial activity of peptides and folding variants from milk proteins [J]. International Dairy Journal, 2006, 16 (11): 1 294-1 305.

[165] Gutierrez J E, Jakobovits L. Capillary electrophoresis of α-lactalbumin in milk powders [J]. Journal of agricultural and food chemistry, 2003, 51 (11): 3 280-3 286.

[166] Gasilova N, Gassner A L, Girault H H. Analysis of major milk whey proteins by immunoaffinity capillary electrophoresis coupled with MALDI-MS [J]. Electrophoresis, 2012, 33 (15): 2 390-2 398.

[167] De Marchi M, Bonfatti V, Cecchinato A, et al. Prediction of protein composition of individual cow milk using mid-infrared spectroscopy [J]. Italian Journal of Animal Science, 2010, 8 (2s): 399-401.

[168] Bobe G, Beitz D C, Freeman A E, et al. Separation and quantification of bovine milk proteins by reversed-phase high-performance liquid chromatography [J]. Journal of Agricultural and Food Chemistry, 1998, 46 (2): 458-463.

[169] Wedholm A, Larsen L B, Lindmark-Månsson H, et al. Effect of protein composition on the cheese-making properties of milk from individual dairy cows [J]. Journal of dairy science, 2006, 89 (9): 3 296-3 305.

[170] Bonfatti V, Cecchinato A, Di Martino G, et al. Effect of κ-casein B relative content in bulk milk κ-casein on Montasio, Asiago, and Caciotta cheese yield using milk of similar protein composition [J]. Journal of dairy science, 2011, 94 (2): 602-613.

[171] Park Y W, Haenlein G F. Handbook of milk of non-bovine mammals [M]. John Wiley & Sons, 2008.

[172] Bu D P, Wang J Q, Dhiman T R, et al. Effectiveness of oils rich in linoleic and linolenic acids to enhance conjugated linoleic acid in milk from dairy cows [J]. Journal of Dairy Science, 2007, 90 (2): 998-1 007.

[173] Yang Z, Liu S, Chen X, et al. Induction of apoptotic cell death and in vivo growth inhibition of human cancer cells by a saturated branched-

chain fatty acid, 13-methyltetradecanoic acid [J]. Cancer research, 2000, 60 (3): 505-509.

[174] Ran-Ressler R R, Sim D, O Donnell-Megaro A M, et al. Branched chain fatty acid content of United States retail cow's milk and implications for dietary intake [J]. Lipids, 2011, 46 (7): 569-576.

[175] Chilliard Y, Martin C, Rouel J, et al. Milk fatty acids in dairy cows fed whole crude linseed, extruded linseed, or linseed oil, and their relationship with methane output [J]. Journal of dairy science, 2009, 92 (10): 5 199-5 211.

[176] R L. Determination of milk odd and branched fatty acids in tropical countries Case study: Cuba [J]. Belgium Ghent University 2011.

[177] Rodríguez E R, Alaejos M, Romero C. Mineral content in goats' milks [J]. Journal of food quality, 2002, 25 (4): 343-358.

[178] Rahimi E. Lead and cadmium concentrations in goat, cow, sheep, and buffalo milks from different regions of Iran [J]. Food chemistry, 2013, 136 (2): 389-391.

[179] Neilson K A, Ali N A, Muralidharan S, et al. Less label, more free: Approaches in label-free quantitative mass spectrometry [J]. Proteomics, 2011, 11 (4): 535-553.

[180] Liao Y, Alvarado R, Phinney B, et al. Proteomic characterization of human milk whey proteins during a twelve-month lactation period [J]. Journal of proteome research, 2011, 10 (4): 1 746-1 754.

[181] Ijsselstijn L, Stoop M P, Stingl C, et al. Comparative study of targeted and label-free mass spectrometry methods for protein quantification [J]. Journal of proteome research, 2013, 12 (4): 2 005-2 011.

[182] Cox J, Mann M. MaxQuant enables high peptide identification rates, individualized ppb-range mass accuracies and proteome-wide protein quantification [J]. Nature biotechnology, 2008, 26 (12): 1 367-1 372.

[183] Fong B Y, Norris C S, Macgibbon A K. Protein and lipid composition of bovine milk－fat－globule membrane [J]. International Dairy Journal, 2007, 17 (4): 275-288.

[184] Zhou L, Chen J, Li Z, et al. Integrated profiling of microRNAs and mRNAs: microRNAs located on Xq27.3 associate with clear cell renal cell carcinoma [J]. PloS one, 2010, 5 (12): e15224.

[185] Anders S, Huber W. Differential expression analysis for sequence count data [J]. Genome biol, 2010, 11 (10): R106.

[186] Wang L, Feng Z, Wang X, et al. DEGseq: an R package for identifying differentially expressed genes from RNA-seq data [J]. Bioinformatics, 2010, 26 (1): 136-138.

[187] Válóczi A, Hornyik C, Varga N, et al. Sensitive and specific detection of microRNAs by northern blot analysis using LNA－modified oligonucleotide probes [J]. Nucleic Acids Research, 2004, 32 (22): e175.

[188] Lu J, Getz G, Miska E A, et al. MicroRNA expression profiles classify human cancers [J]. nature, 2005, 435 (7043): 834-838.

[189] Fichtlscherer S, De Rosa S, Fox H, et al. Circulating microRNAs in patients with coronary artery disease [J]. Circulation research, 2010, 107 (5): 677-684.

[190] Zhao H, Shen J, Medico L, et al. A pilot study of circulating miRNAs as potential biomarkers of early stage breast cancer [J]. PLoS One, 2010, 5 (10): e13735.

[191] Chen C, Ridzon D A, Broomer A J, et al. Real-time quantification of microRNAs by stem-loop RT-PCR [J]. Nucleic acids research, 2005, 33 (20): e179.

[192] Yang H, Schmuke J J, Flagg L M, et al. A novel real-time polymerase chain reaction method for high throughput quantification of small regulatory RNAs [J]. Plant biotechnology journal, 2009, 7 (7):

621-630.

[193] Benes V, Castoldi M. Expression profiling of microRNA using real-time quantitative PCR, how to use it and what is available [J]. Methods, 2010, 50 (4): 244-249.

[194] Hu Z, Chen X, Zhao Y, et al. Serum MicroRNA signatures identified in a genome-wide serum MicroRNA expression profiling predict survival of non-small-cell lung cancer [J]. Journal of Clinical Oncology, 2010, 28 (10): 1 721-1 726.

[195] Wu Q, Lu Z, Li H, et al. Next-generation sequencing of microRNAs for breast cancer detection [J]. BioMed Research International, 2011: 597145.